IMAGES
*of America*

# CRAFTSMAN FARMS

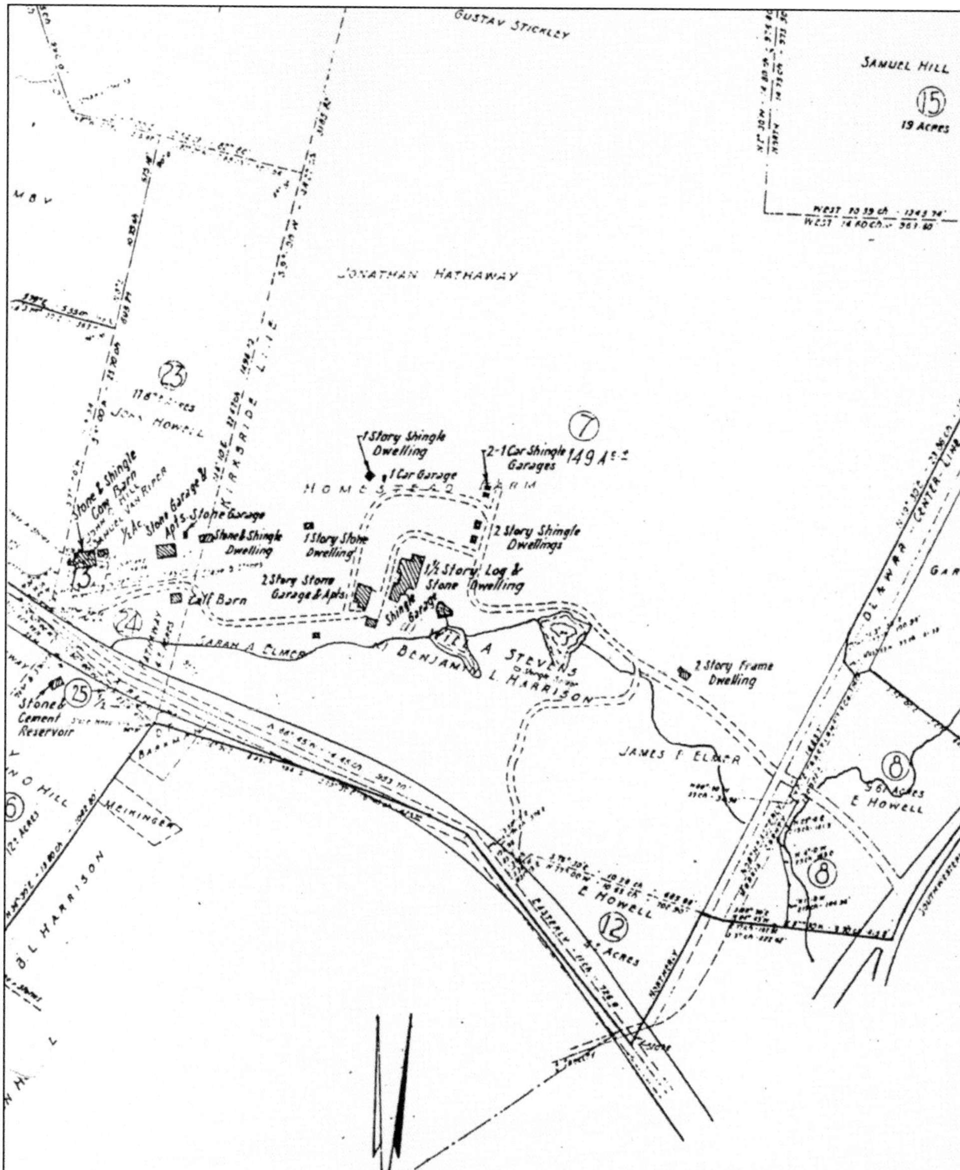


This modified survey of Craftsman Farms after the construction of New Jersey State Highway 10 shows Craftsman Farms and the locations of the buildings, streams, ponds, and internal roads. Gustav Stickley's name can be seen at the top. The names of some of the previous property owners are also listed on the map. (Courtesy Morris County, Deeds and Mortgages.)

**ON THE COVER:** This photograph is a detail of a larger view to the east at Craftsman Farms around 1915. The northern end of Gustav Stickley's log house is shown with the dining room's casement windows open to bring in fresh country air. Foundation plantings and the lawn show how Stickley intentionally contrasted the home's more formal landscape with the rough and more naturalistic landscape in the nearby ravine. The pale white line that can be seen in the distance is the railroad track bed, which marks the estate's eastern border. (Courtesy the Stickley Museum at Craftsman Farms.)

IMAGES
*of America*

# CRAFTSMAN FARMS

Heather E. Stivison

ARCADIA
PUBLISHING

ISBN 978-1-4671-2205-4

Published by Arcadia Publishing
Charleston, South Carolina

Printed in the United States of America

Library of Congress Control Number: 2013957043

For all general information, please contact Arcadia Publishing:
Telephone 843-853-2070
Fax 843-853-0044
E-mail sales@arcadiapublishing.com
For customer service and orders:
Toll-Free 1-888-313-2665

Visit us on the Internet at www.arcadiapublishing.com

*For my eternally optimistic husband, Douglas*

# Contents

# ACKNOWLEDGMENTS

Craftsman Farms would not exist today without the dedicated efforts of many individuals and organizations. In the 1980s, when the property was threatened with destruction, a grassroots effort was launched to prevent its conversion into townhouses. In April 1989, the mayor and town council of Parsippany–Troy Hills responded by using eminent domain and Green Acres and Open Space funds to purchase Craftsman Farms. Grants from Morris County Historic Preservation Trust, the New Jersey Historical Commission, and the New Jersey Historic Trust have also played major roles in the preservation of the site. Since 1989, the preservation of Craftsman Farms has been handled by the nonprofit Craftsman Farms Foundation, which operates the property as the Stickley Museum at Craftsman Farms. A huge debt of gratitude is owed to the foundation's trustees, both past and present, and to the numerous volunteers who continue to give their time and talents to preserve and interpret the property. I am proud to publicly thank the dedicated staff of the Stickley Museum, who were my colleagues for nearly seven years, for the extraordinary work they do each and every day.

In addition to people who have helped preserve and restore Craftsman Farms, I thank those individuals who have helped make this book possible, particularly the members of the Stickley family and the Farny family who allowed their family photographs to become part of this historical record. I am especially grateful to Evelyn Wadkins and her parents, Ethel and Michael Farny, who generously shared their time and family memories. A debt of gratitude goes to David Cathers, who not only is the preeminent authority on Gustav Stickley but whose friendship during my years at the Stickley Museum was truly inspirational. A big thank you goes to Peter Mars, vice president of the Craftsman Farms Foundation, who provided invaluable assistance, and to the registrar, Bernadette Rubbo, who cheerfully gathered numerous images from the museum's archives. I am grateful to my editor, Katie McAlpin, for providing unflagging enthusiasm and expert guidance throughout the process. And finally, I thank my husband, Doug, for his unwavering support throughout this entire project. Unless otherwise noted, images are courtesy of the Stickley Museum at Craftsman Farms.

# INTRODUCTION

Craftsman Farms was created in the early 1900s by noted designer, furniture manufacturer, publisher, and leader of the American Arts and Crafts movement Gustav Stickley. His design firm is credited with creating more than 220 house designs, but he had never before designed a home for his own use. Here at Craftsman Farms, he finally designed and built a dream home for himself and his family. He rhapsodized about that process in the October 1908 issue of his successful magazine, *The Craftsman*:

> I never before realized how much pleasure was to be found in the building of a dwelling that as completely expressed one's own taste and individuality as the painting of a picture or the writing of a book. In fact, I can think of no creative work that is so absorbingly delightful as this creation of a home to live in for the rest of one's life.

Gustav Stickley's country property, Craftsman Farms, has often been called the cradle of the green movement in New Jersey. Stickley began creating this idealistic and rustic retreat in 1908—a time in which much of American society faced concerns very much like those of our own time. Today, although we love the rapidly changing technology that makes our lives easier, we also feel some uneasiness with the environmental costs that go hand-in-hand with our high-tech lives. We often find ourselves feeling out of touch with the natural world and yearn for a life that is simpler, more peaceful, and more in touch with nature. An increasing number of us say we need to be "greener" and seek healthier lifestyles, buy farm shares, and support "farm-to-table" restaurants. We are concerned about the roller coaster of our nation's economy, high unemployment, and people who are stuck working for large corporations in jobs they find meaningless. This present-day mood has fueled a surge in small independent businesses, such as those that are crowdfunded through sites like Kickstarter, and a growth in independent craftspeople creating work they find more satisfying and selling it on sites like Etsy.

Similarly, in Stickley's day, a rapid increase in manufacturing new products also made lives easier, but that too came with a price. Growing urbanization and powerful business monopolies led to a new type of poverty, joblessness, and poor working conditions. The nation's farmland, rural areas, and wilderness all suffered. Suggestions arose about healthier lifestyles and getting back to nature in the early 1900s, just as they are arising today. Pres. Theodore Roosevelt's progressive agenda on environmental conservation and social justice was developed in response to these problems. And this progressive agenda was clearly aligned with Gustav Stickley's goals in his creation of Craftsman Farms.

There are indications that, for quite some time, Stickley had idealistic thoughts about creating a self-sufficient rural farm and handcraft community. A lengthy review of a book entitled *Back to the Soil* appears in the November 1901 issue of Gustav Stickley's magazine *The Craftsman*. Though constructed as a novel, its author, Unitarian minister Bradley Gilman, was clearly advocating for creating small farming and craft villages as a healthier and more rewarding lifestyle. Gilman indicated these small villages should be funded and "headed by the self-made capitalist, who to the instincts and desires of a money-getter joins the warm heart of a friend of humanity." On a 1904 trip to California, Stickley and journalist James Wharton James make plans for just such a community on the West Coast. Though that project never materialized, Stickley continued to consider the concept.

In 1905–1906, with his business prospering and his national fame continuing to grow, Stickley moved his publishing business from Syracuse, New York, to New York City and established his Craftsman showrooms on Thirty-Fourth Street. Soon, his friend, Hearst newspaper cartoonist Homer Davenport, invited Stickley to visit Davenport's newly completed country estate, Red Gables, which was located in bucolic Morris Plains, New Jersey.

With its easy commute to New York City and spectacular country estates, the area struck a chord with Stickley, and by the spring of 1908, his account books show weekly train fares between New York and Morris Plains. In June of that year, Stickley purchased 30 acres of land adjoining Davenport's Red Gables in Morris Plains. Two more parcels of land were purchased in July, including one known locally as Garrigus Farm.

In August of that same summer, Pres. Theodore Roosevelt created the Country Life Commission, which he said was "to direct the attention of the nation to the problems of the farm, and also for securing the necessary knowledge of the actual conditions of life in the open country." Clearly, Stickley and Roosevelt were aligned in their shared reverence for the landscape in America and a concern for the well-being of its citizens. In the October 1908 issue of *The Craftsman*, Stickley remarked, "the principles which, from the first number of the magazine, we have so persistently advocated, are now shaping not only the best thought but the best legislation of today."

In September 1908, Stickley purchased additional property to the north of Whatnong Mountain and Mount Pleasant Turnpike in Morris Plains. By March of the following year, he had spent some $10,000 on land acquisition. The property to be known as Craftsman Farms would eventually total 650 acres.

Stickley's magazine, *The Craftsman*, featured the dreams, passions, and ideals of the American Arts and Crafts movement—and of Stickley himself. Soon, his optimism and excitement about Craftsman Farms and what he saw as the "prevailing spirit of the age" filled the magazine.

In the October 1908 issue of *The Craftsman*, he wrote with great fervor about his plans for the following year:

> We have set our faces toward the growing light and our feet are planted on the rising ground, and the most encouraging part of it all is that the effort toward better things is not confined to governments, societies or prominent and powerful, individuals, but the great body of the people is beginning to realize that life and work need not be oppressive, that education need not be inadequate as a training for all practical affairs, and that the essence of happiness and prosperity does not necessarily lie in the feverish pursuit of the means wherewith to gratify artificial tastes and needs.

That same issue included an article entitled "The Craftsman's House: A Practical Application of All the Theories of Home Building Advocated in this Magazine," which was a very detailed description of the home Stickley planned to build for himself at Craftsman Farms. His passion for the place and his project was apparent throughout.

> The estate of Craftsman Farms was apparently planned by nature for just the use to which I intend to put it. It has heavily wooded hills, little wandering brooks, low-lying meadows and plenty of garden and orchard land, and the site of my own house is on a natural terrace or plateau half way up the highest hill. . . . Beyond this and over the tops of the low-lying hills around one looks straight out to the line of the Orange Mountains in the misty blue distance. . . . The picturesque possibilities of the place are so great that it is a

> marvel that they have never been seen or developed, especially as this part of Jersey is the home of beautiful country places. There is a friendliness about the natural conformation of the land that makes it seem homelike before one stone is laid upon another or one bit of underbrush is cleared away.

Never short of ideas and theories, Stickley's overarching vision for Craftsman Farms was as grand a vision as anyone could imagine. Early in the Craftsman Farms planning process, he announced he would create Craftsman Village, which was to be an idealistic community of artisans and small farmers, just as he and his friend, James Wharton James, had considered in 1904. The October 1908 issue of his magazine reported that this big idea, complete with a school, was about to become a reality.

> For the first time I am applying to my own house, and working out in practical detail, all the theories which so far I have applied only to the houses of other people. For this reason I am publishing the perspective drawings, floor plans and details of construction of the house I am building at "Craftsman Farms," an estate I have recently purchased near Morris Plains, New Jersey, upon which I purpose to make my home and where I am preparing to establish a school for the definite working out of the theory I have so long held of reviving practical and profitable handicrafts in connection with small farming carried on by modern methods of intensive agriculture.

Gustav Stickley had been so successful in his rise to fame that he must have assumed all his dreams would come true. He was also a great salesman who often described his magnificent ideas and plans as if they were facts. But his vision for Craftsman Farms was truly a massive undertaking and one wonders how any man could create what he described. Stickley was also a man who was responsible for a massive design business, a large furniture manufacturing business, and a showroom in New York City, as well as being a publisher of note. How could he possibly devote the time, energy, and money to complete Craftsman Farms as he envisioned it?

The plans for Craftsman Village were abandoned and evolved into plans for a school for boys, which later magazine articles called the Gustav Stickley School for Citizenship. Educational reform was a major issue of the time, and Stickley weighed in frequently, with thoughtful articles appearing in his publications as early as January 1902. This was only a few years after John Dewey opened his famous Laboratory School at the University of Chicago. Dewey's school stressed learning by doing—a theory embraced by Stickley. The September 1908 issues of *The Craftsman* explores this concept in an article entitled "A Visit to Craftsman Farms: The Study of an Educational Ideal." In this article, an imaginary visitor learns from Stickley (described as the host) his overall plans for Craftsman Farms.

Meanwhile, farming was ramping up successfully at Craftsman Farms. Construction was underway on the first cottages, and by the spring of 1910, these cottages were completed. In July, his wife, Eda, and daughters Mildred, Marion, Hazel, and Ruth left their home in Syracuse, New York, and settled into one of the cottages.

Construction also began on the log house, which was originally intended to be the clubhouse for the school. Stickley continued to publicly discuss plans for his school, although during construction plans were changed for the clubhouse. The log house was redesigned to become the Stickley family's home. Elements were adapted from the residence he had originally planned to build. In the process of this transformation, Stickley began creating what was to become one of the most significant buildings in the American Arts and Crafts movement.

In a building style that undoubtedly resonates with today's green movement, the massive log house was built from stones that were dug on the property and logs hewn from trees in the surrounding woods. The house was completed in 1911, and the family took up residence in that same fall. Its interior was described in the November 1911 issue of *The Craftsman* as reminiscent "of the forest—brown and green with the glint of sunshine through the leaves, suggested by the gold of the windows and the gleam of copper in the hearth hoods, the door latches, and the vases and bowls on the bookcases and tables."

Craftsman Farms was soon well on its way to achieving Stickley's goal of self-sufficiency. Orchards of apples, cherries, peaches, and pears were thriving to the west of the log house. Most of these orchards must have already been established in some of the parcels of farmland Stickley purchased and were likely carefully considered in the placement of the log house and other buildings. A grape arbor flourished to the north, and soon the property was producing vegetables in the fields. Milk was plentiful from the Holstein cows that enjoyed free-range grazing. Healthful wholesome milk was assured with his farming practices and newer ideas such as the impressive barn—complete with a Craftsman fireplace to help guard against the winter's cold. The barn also had tall windows for healthful ventilation and window screens for protection against flies. By 1913, Stickley had managed to create today's version of a farm-to-table restaurant. Craftsman Farms was providing healthy eggs, milk, and produce to his newly opened Craftsman Restaurant in New York City.

The multi-page menu to Stickley's Craftsman Restaurant was full of text about Stickley's ideals and photographs of Craftsman Farms. The menu proudly proclaimed a message very much like today's concept of farm-to-table dining.

> My theory about a restaurant is that to be the right sort of an eating place it must be closely related to its source of supplies. In no other way can it be fully hygienic and satisfactory. The farm products must come direct from producer to consumer. To test this theory, I decided to combine the Craftsman Restaurant with the Craftsman Farms, and to bring my farm products to my restaurant table in New York without the delay and the consequent deterioration of food inevitable in many handlings by wholesale and retail dealers.

Craftsman Farms demonstrated Stickley's firm commitment to working in harmony with the natural world and to a simpler life. Though changing economic times caused his grand experiment to be cut short, the ideas still resonate with people today. However brief his years were at Craftsman Farms, the place and ideas that formed it still have much to teach people today. This book invites the reader to explore Craftsman Farms through images and words and experience some of the romance and beauty that can be found there to this very day.

# *One*

# Gustave Stickley Pursues His Dreams

On March 9, 1858, in a place known today as Osceola, Wisconsin, Barbara and Leopold Stoeckel welcomed their first son, Gustave, into the world. The Stoeckels, first-generation German immigrants whose family name meant "little stick," soon Americanized their name to be the now familiar "Stickley." Wisconsin had just joined the union as a state in 1848 and offered fresh possibilities for a better life. And so, like many German Americans in the 1850s, Gustave's parents settled in the frontier to establish a 40-acre homestead with hopes to achieve that better life for their family.

Despite their hopes, life was not easy for them. Leopold worked as both a farmer and stonemason, but they still struggled financially. Of their 11 children, only 9 survived to adulthood. Young Gustave's formative years included doing rugged farm work as well as helping his father with the stone masonry. And, like many children at the time, Gustave left school by the age of 12.

Times were indeed hard, and Leopold and Barbara separated briefly. Leopold returned for a short period, and the family moved to Stillwater, Minnesota, in the hopes of finding more work for Leopold. Those hopes did not come to fruition, and the couple separated again. When Gustave was about 18 years old, Barbara and her children settled in Lanesboro, Pennsylvania, where her brother, Jacob Schlager, operated a chair factory.

Schlager hired his nephew, Gustave, to work in the factory, where he learned a great deal about the furniture business and developed a real love for working with wood. Gustave became the factory foreman in just a few years and must have begun to develop the skills to manage people that would serve him well in the future. A keen observer of life, he certainly would have also noted the stark difference between the hardscrabble life he had known on the farm and the life of the prosperous furniture factory owner. These experiences fueled the ambitions that set him on the course for his future.

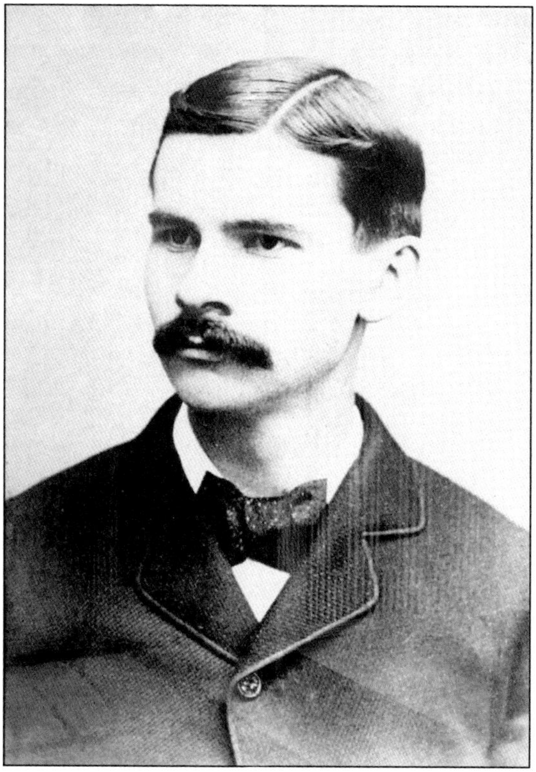

This portrait of Gustave Stickley is believed to be from 1875. It was at this time that he began to work in a furniture factory in Brandt, Pennsylvania, that was operated by Henry William Brandt and Stickley's uncle Jacob Schlager. By 1883, Stickley and his brothers Albert and Charles had established their own furniture manufacturing business in Susquehanna, Pennsylvania, called Stickley Brothers & Company. (Courtesy the Stickley Museum at Craftsman Farms, Capt. Peter Wiles Sr. Collection)

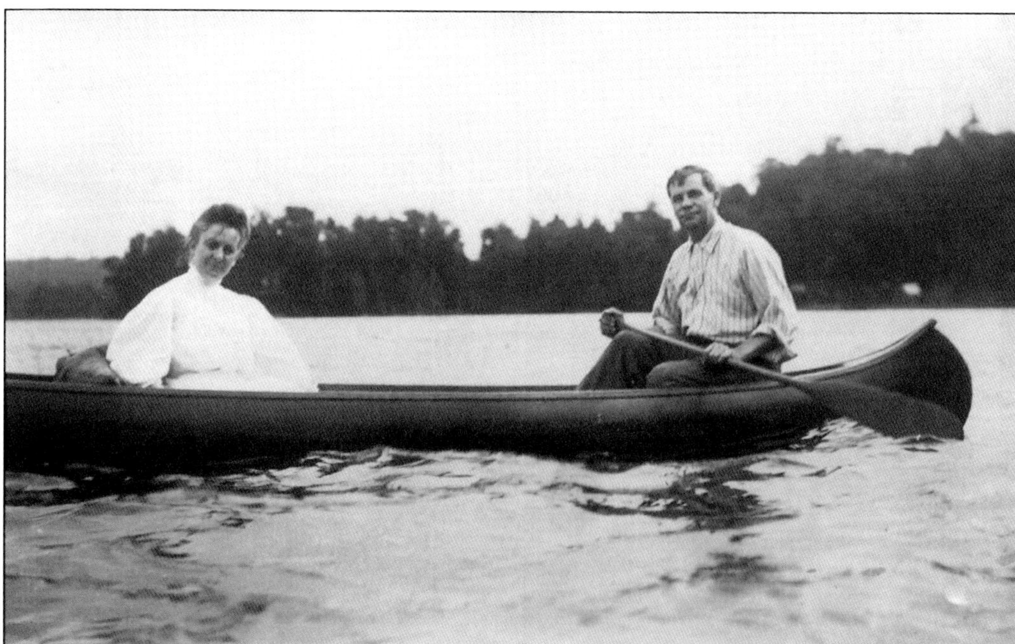

Gustave married Eda Simmons in 1883 when she was 23 and he was 25. In a rare photograph of the two of them together, Eda and Gustave are seen here canoeing in the early 1900s. In the 1890s, the family began vacationing in the Adirondacks, where Gustave must have encountered the large rustic log cabins that would later inspire him at Craftsman Farms. (Courtesy Cynthia McGinn.)

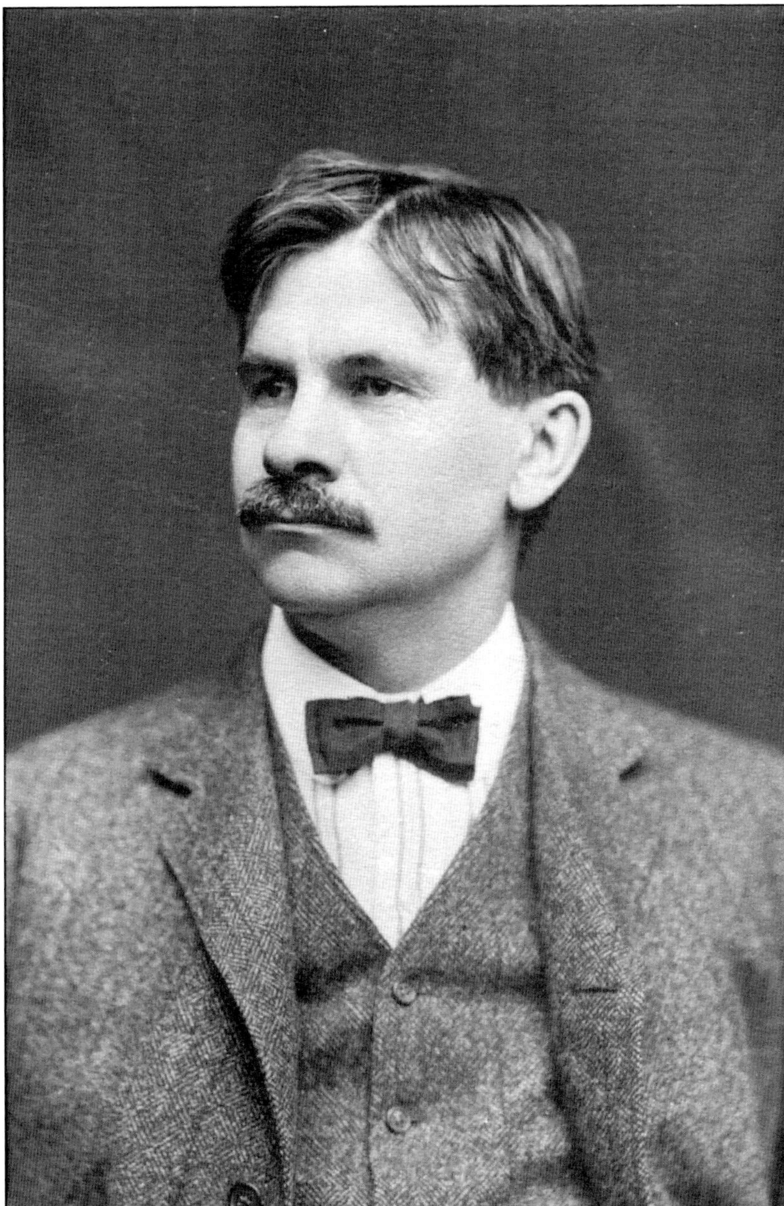

In 1888, Gustave Stickley resigned from Stickley Brothers & Company to partner with furniture salesman Elgin Simonds in a new furniture manufacturing business, the Stickley and Simonds Company. This thriving business venture soon helped build Stickley's wealth as well as his knowledge of both the business world and of new concepts in design. He traveled to Europe and gained inspiration from the styles he saw there. After a decade of partnership, Stickley and Simonds were clearly differing on what direction the business should take. Stickley asserted that he no longer wanted to manufacture historic reproductions and adaptations, suggesting they were soulless and lacking in spirit. In 1898, the year this portrait was taken, he took the somewhat ruthless steps of buying out Simonds's shares of the company, meeting with the board of directors, and arranging for the dissolution of the Stickley and Simonds Company. Within months, the Gustave Stickley Company was formed.

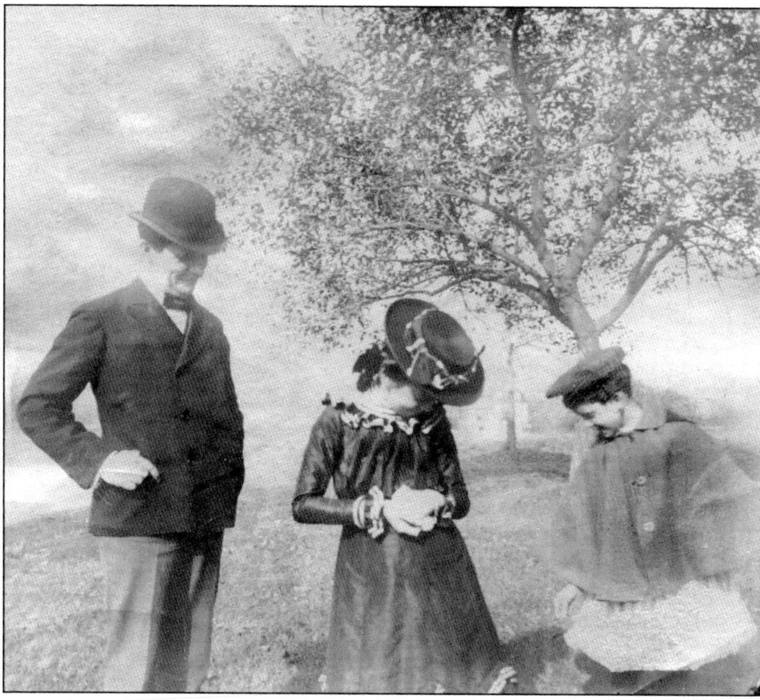

While Gustave's business was growing, his family was also growing. By 1897, he and Eda were the proud parents of six children: Barbara, Mildred, Hazel, Marion, Gustav Jr., and Ruth. His prosperity was apparent as the family moved from Auburn, New York, to a house on Walnut Street in Syracuse. He is shown here in the late 1890s with daughters Mildred (left) and Hazel (right).

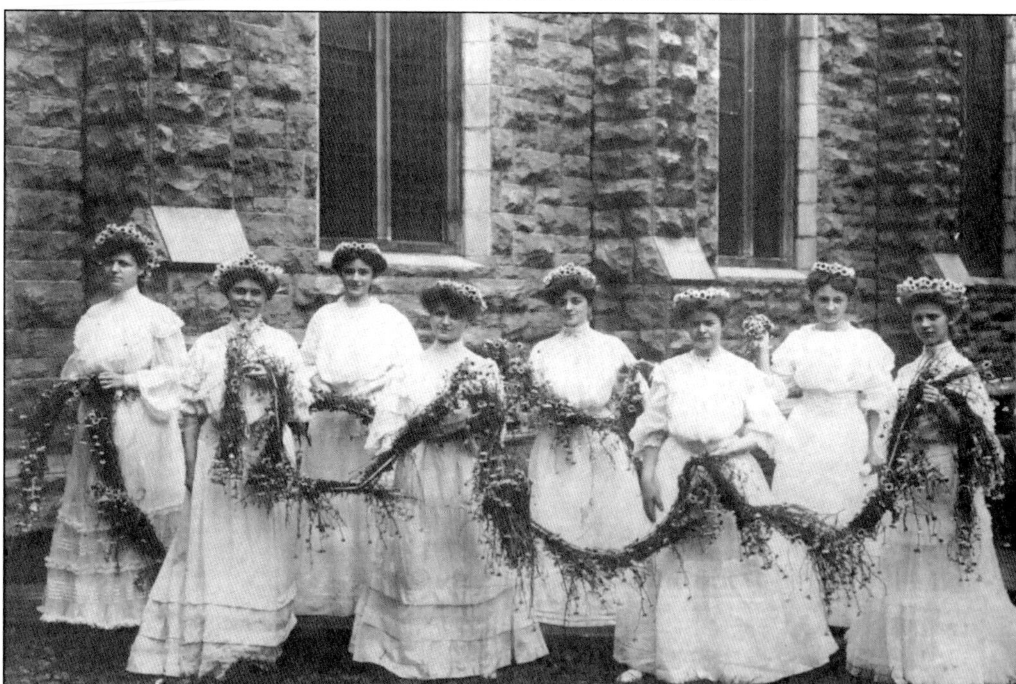

In Syracuse, the Stickley daughters enjoyed the niceties their father's success provided and the social life the community had to offer. In this photograph, the eldest daughter, Barbara (far right), is shown with friends wearing classic white "lawn" or "lingerie" dresses, possibly at a May Day celebration. Gustave's prestige continued to rise, and in 1900, the family moved into a spacious Colonial Revival home on Columbus Avenue. (Courtesy Barbara Fuldner.)

Stickley introduced experimental pieces and his early Arts and Crafts style. Talented new designers Henry Wilkinson and LaMont Warner were hired. In 1900, the Tobey Furniture Company distributed Stickley's furniture and was the first to advertise it as "Arts and Crafts." Stickley leased space in the Crouse Stables, Syracuse, shown here as it appeared in 1904, when Stickley renamed it the Craftsman Building.

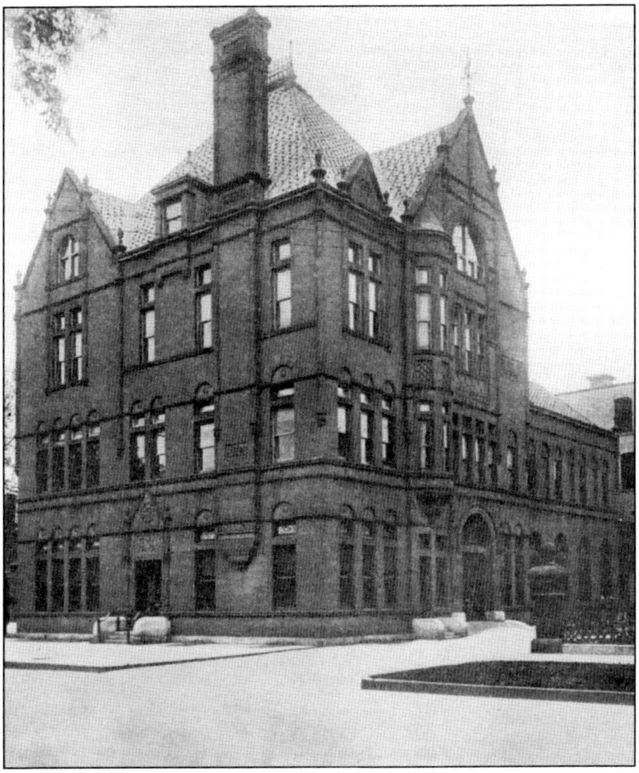

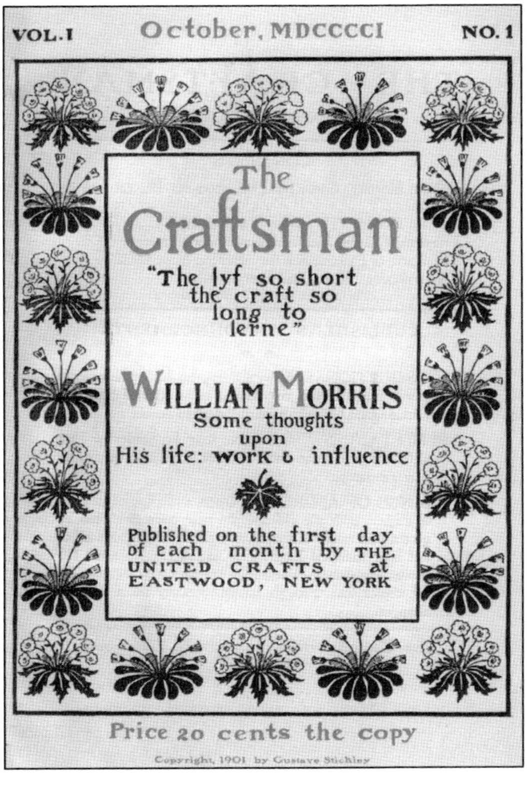

VOL. I October, MDCCCCI NO. 1

The Craftsman

"The lyf so short the craft so long to lerne"

WILLIAM MORRIS
Some thoughts upon His life: work & influence

Published on the first day of each month by THE UNITED CRAFTS at EASTWOOD, NEW YORK

Price 20 cents the copy

Copyright, 1901 by Gustave Stickley

In October 1901, Stickley introduced *The Craftsman* magazine, which would become one of his most lasting legacies. Scholar David Cathers notes that Eda and the children assembled the early issues by hand at the kitchen table. This first issue was dedicated to British Arts and Crafts icon William Morris. The cover shows that Stickley's company had been rebranded as the United Crafts.

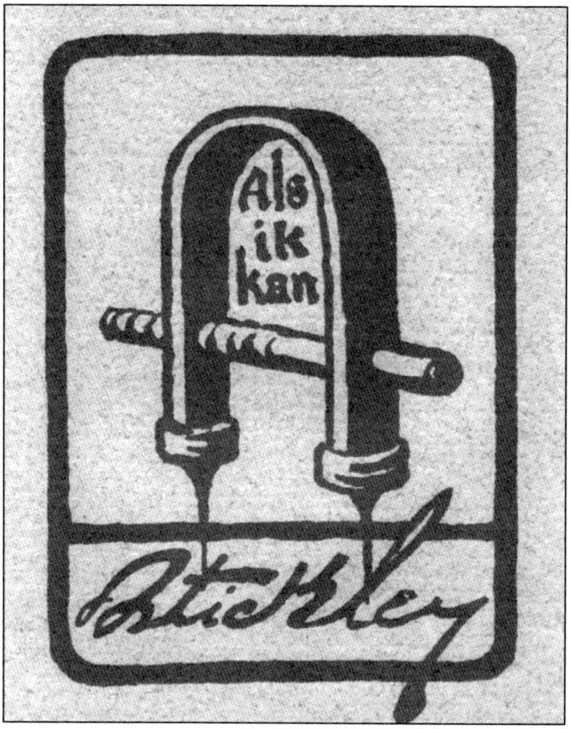


Stickley began using his now famous motto "*Als Ik Kan*" in 1901. He referenced Flemish painter Jan Van Eyck and William Morris, who had used it in the past. Stickley used it to mean "all I can" and soon paired it with a drawing of a tool known as a joiner's compass, shown here in one of its later variations.

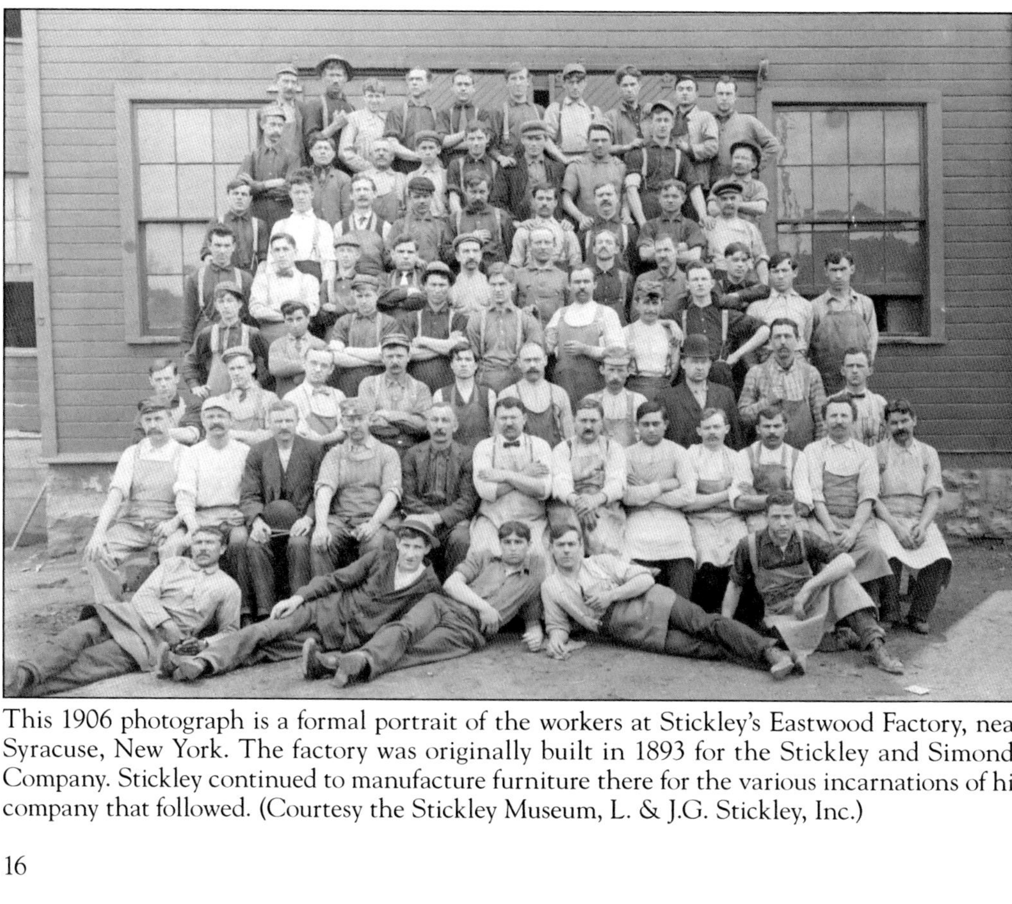

This 1906 photograph is a formal portrait of the workers at Stickley's Eastwood Factory, near Syracuse, New York. The factory was originally built in 1893 for the Stickley and Simonds Company. Stickley continued to manufacture furniture there for the various incarnations of his company that followed. (Courtesy the Stickley Museum, L. & J.G. Stickley, Inc.)

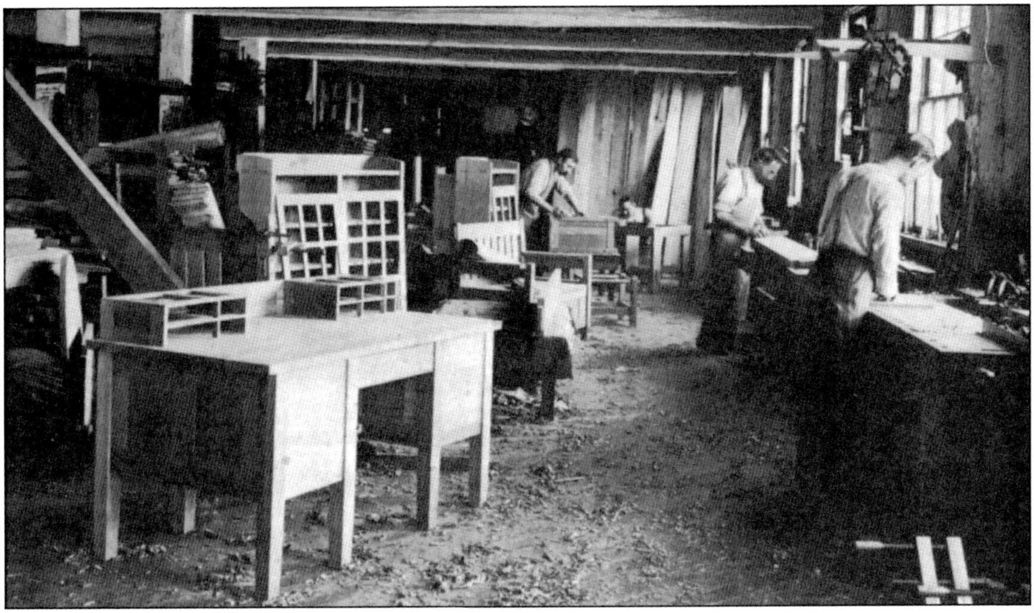

The October 1902 issue of *The Craftsman* featured an article entitled "A Visit to the Workshops of the United Crafts at Eastwood, New York." In it, readers could see the deeply engaged artisans busily crafting their wooden products in a pleasant rural setting. This sent a message that was very much in keeping with the Arts and Crafts vision of elevating the work of one's hands. Stickley was clearly setting his craftsmen apart from the usual urban factory workers. Instead, he positioned his brand as one that offered a variety of products designed with integrity and manufactured in a socially conscious integration of handcraft and machine work. This theme is apparent in a later image of needle workers in his textile workshop.

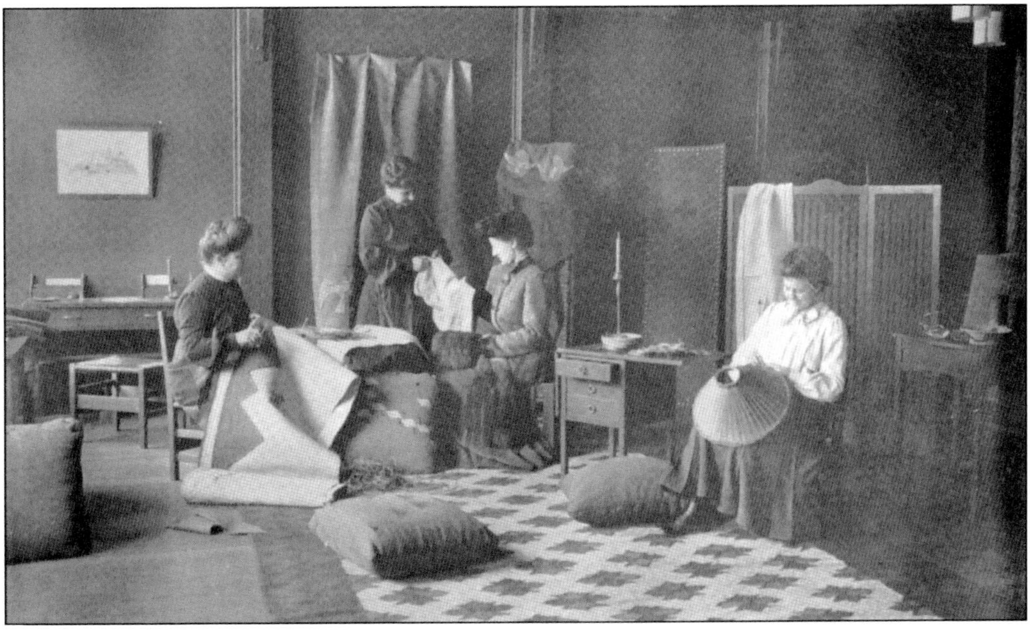

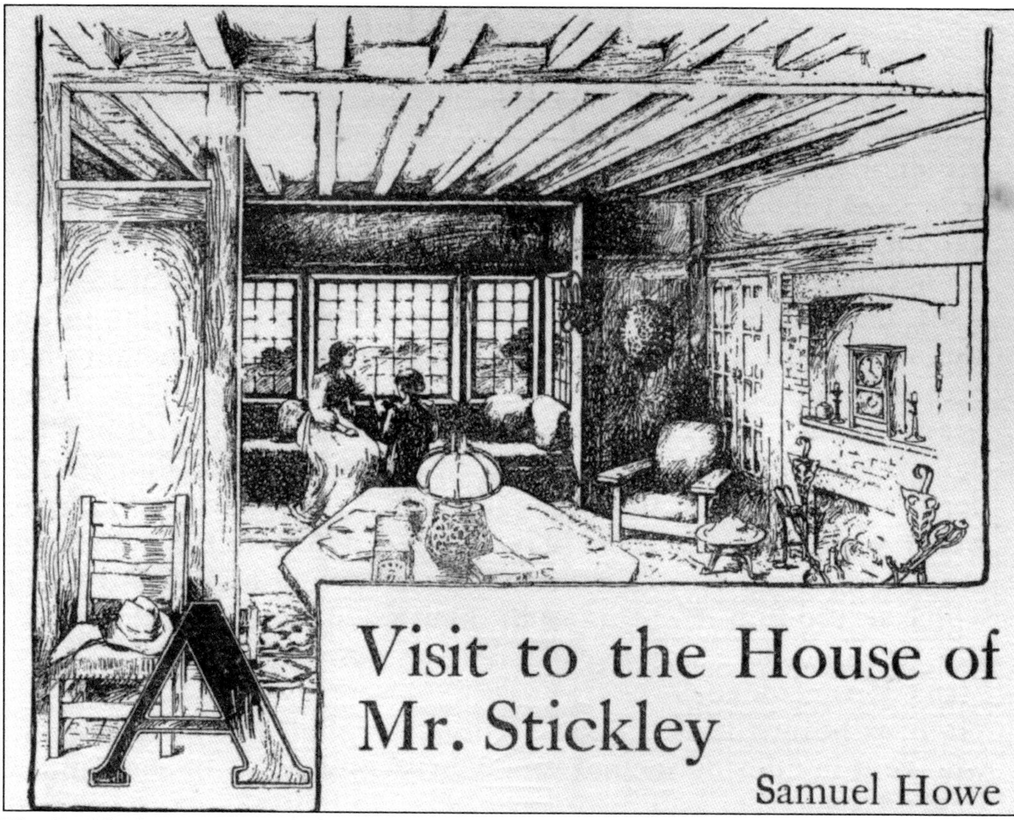


The Stickley home was nearly destroyed in a fire on Christmas Eve, 1901. Gustave made this disaster into an opportunity to create an interior environment that would fully realize the ethos of the Craftsman home. The December 1902 *Craftsman* featured the refurbished interiors in an article entitled "A Visit to the House of Mr. Stickley," which included some of the drawings shown here. Their peaceful, honest, and simple appearance was in stark contrast to existing fussy and lavish Victorian tastes. This was a remarkable period for Stickley's growing enterprises, and he was thriving in his role as publisher and tastemaker. But once again, the difficulties of strong, single-minded personalities seemed to be at play. In 1901, his brother Leopold resigned as factory foreman, and in 1902, he partnered with his other brother, John George, to form the L. & J.G. Stickley Company manufacturing similar products in nearby Fayetteville, New York. (Both, courtesy Barbara Fuldner.)

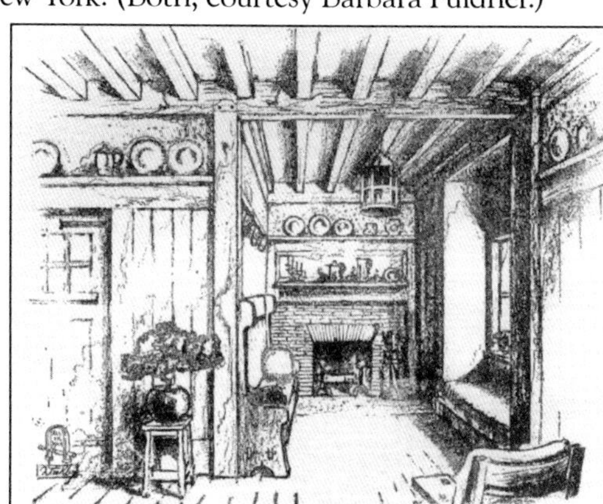

Inglenook, Dining-room

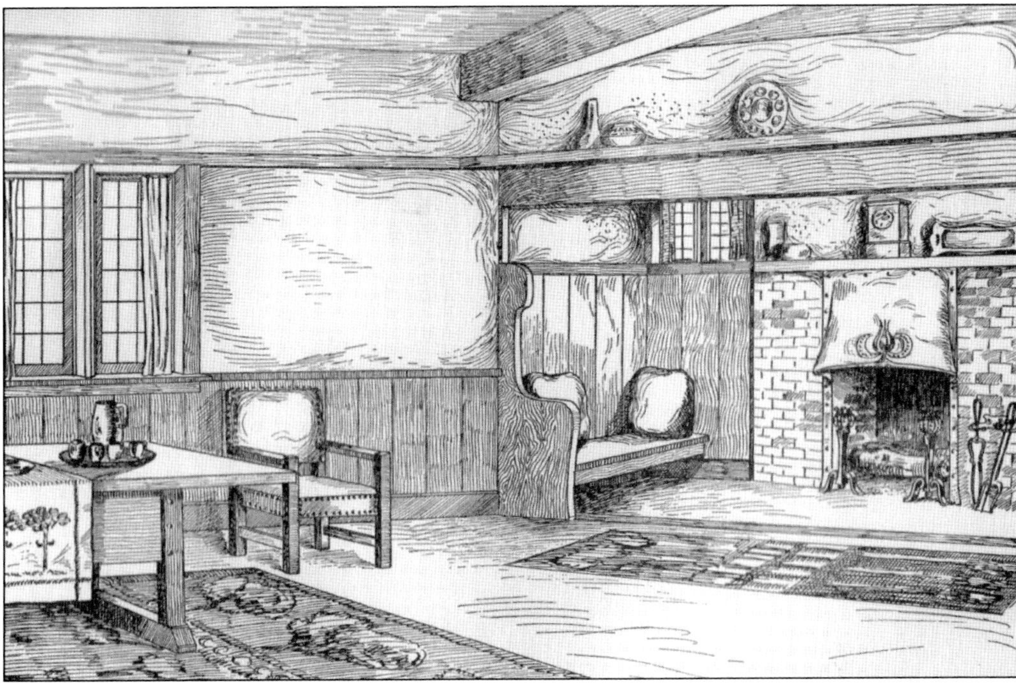

Stickley expanded his corporate identity beyond furniture into an identity with a significant ideology—full of dreams and the ideals of visionaries like William Morris and John Ruskin. Visions for ideal home environments were offered as free house plans for subscribers. The May 1903 issue of *The Craftsman* included the first complete home plan labeled "Craftsman." It featured the dining room shown here.

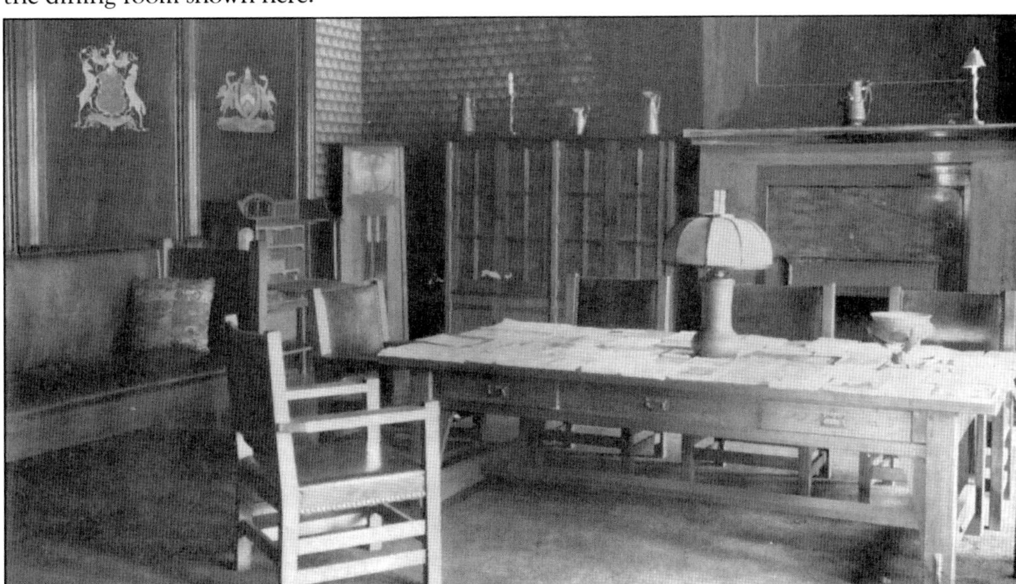

Thanks to his editors Irene Sargent and later Mary Fanton Roberts, Stickley moved from the edges of the Arts and Crafts movement to a major force at the heart and soul of the movement. *The Craftsman* was in the thick of the societal, educational, and economic reforms of the period. A staged setting of the editorial offices is seen in this 1903 photograph.

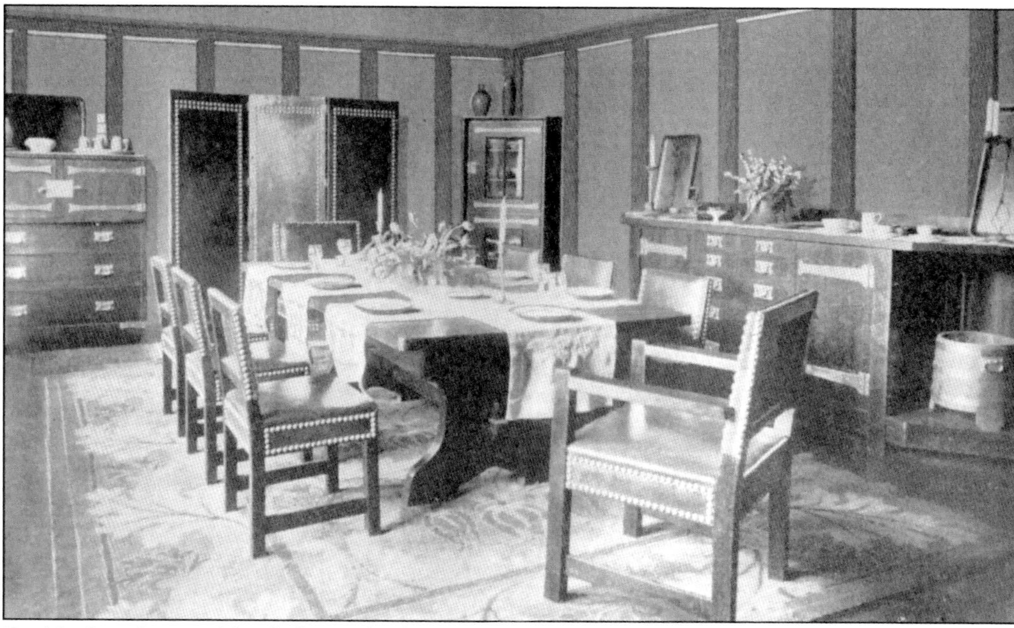

Stickley's endeavors blended moral reform, idealism, and honest design, stirred up with personal ambition. His Craftsman Building was the site of lectures, events, and an Arts and Crafts exhibition, which included the dining room shown here. He further refined his image by changing the name United Crafts to the Craftsman Workshops and changing his own name to eliminate the "e" in Gustave.

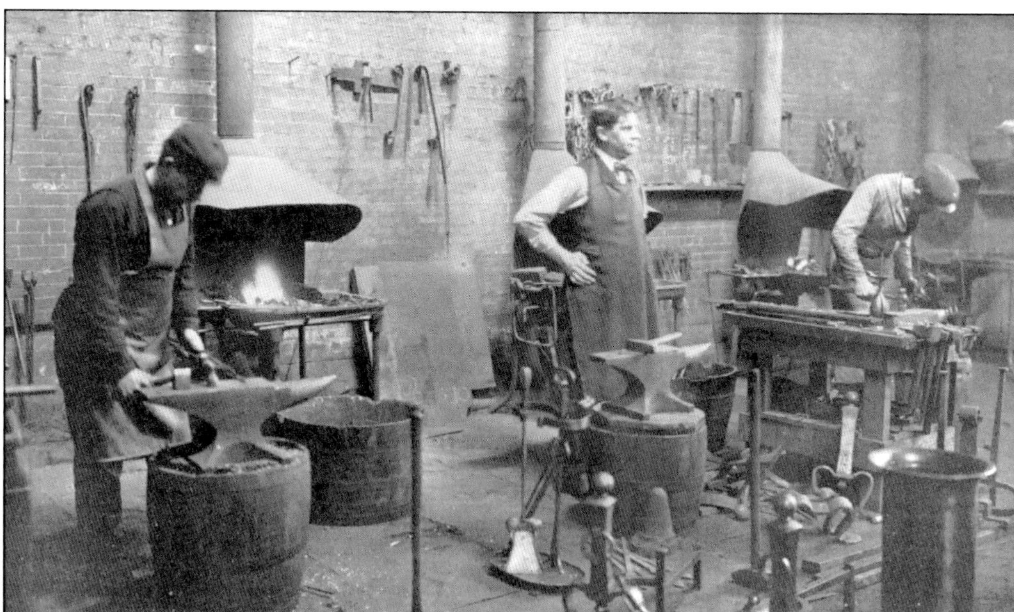

Handcrafted home furnishings made of metal and textiles were designed specifically to complement Craftsman furniture and became a growing part of the business. These items offered customers a way to create complete and harmonious interior environments. Stickley (center) can be seen here posing between two of his metalworkers in this staged photograph that appeared in *What is Wrought in the Craftsman Workshops* in 1904.

On the cover of the January 1904 issue of *The Craftsman* was an image of a California mission, which was to accompany a feature article by George Wharton James on the topic of California mission design. In the spring of 1904, Gustav traveled to California with his wife and two elder daughters. Author George Wharton James, who, like Stickley, was energetic and multifaceted, was their guide. House plans, for the house shown here, appeared in the magazine later that year. Its design was clearly influenced by Stickley's trip to California. During this seven-week California trip, Stickley and James also discussed a desire to create an idealistic place to live "a simple life." These thoughts would become the seeds for his plan for Craftsman Farms.

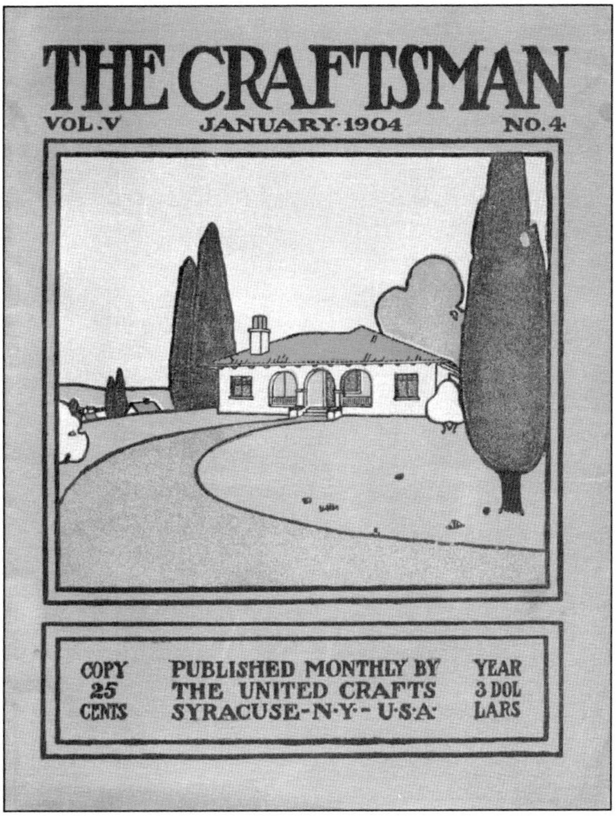


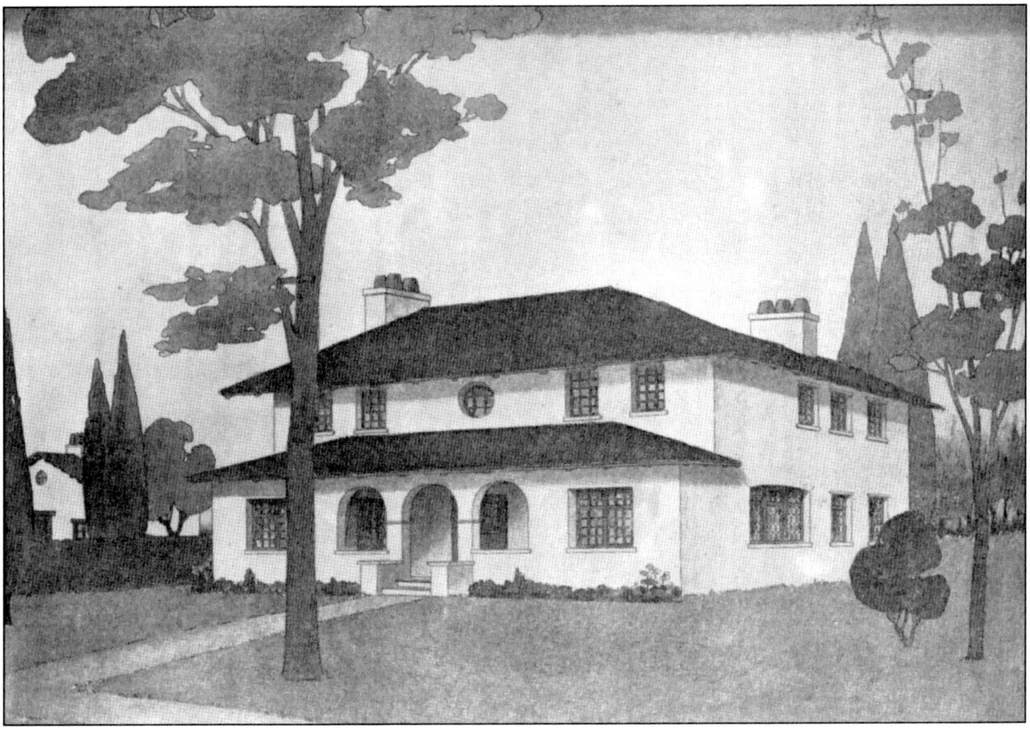

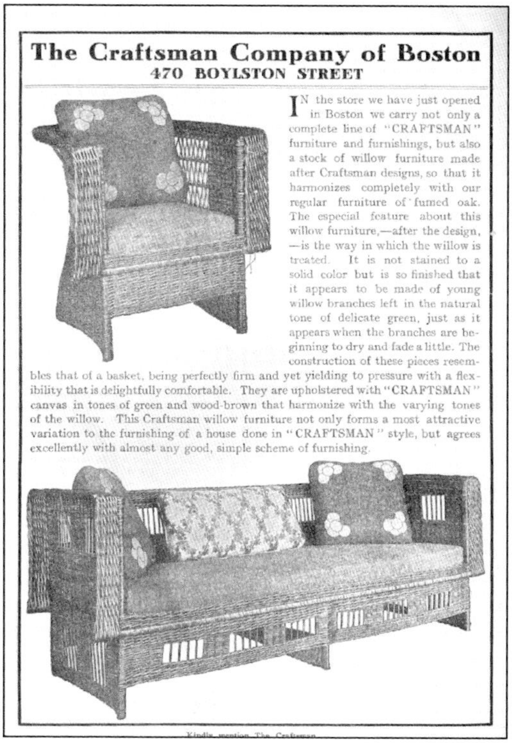


In addition to the growing national awareness of *The Craftsman*, Stickley actively cultivated a national presence in his retail and wholesale furnishings business. His furniture was exhibited in the Boston Mechanics Fair in 1902 and in a decorative arts exhibition at the Art Institute of Chicago in 1903. In 1908, he opened a large new retail store at 470 Boylston Street in Boston. His line of willow furniture was featured in this advertisement for the Boston store. Rugs, embroidered linens, lighting, and accessories can be seen in the display featured in this interior photograph of his Boston retail store. (Below, courtesy Cynthia McGinn.)

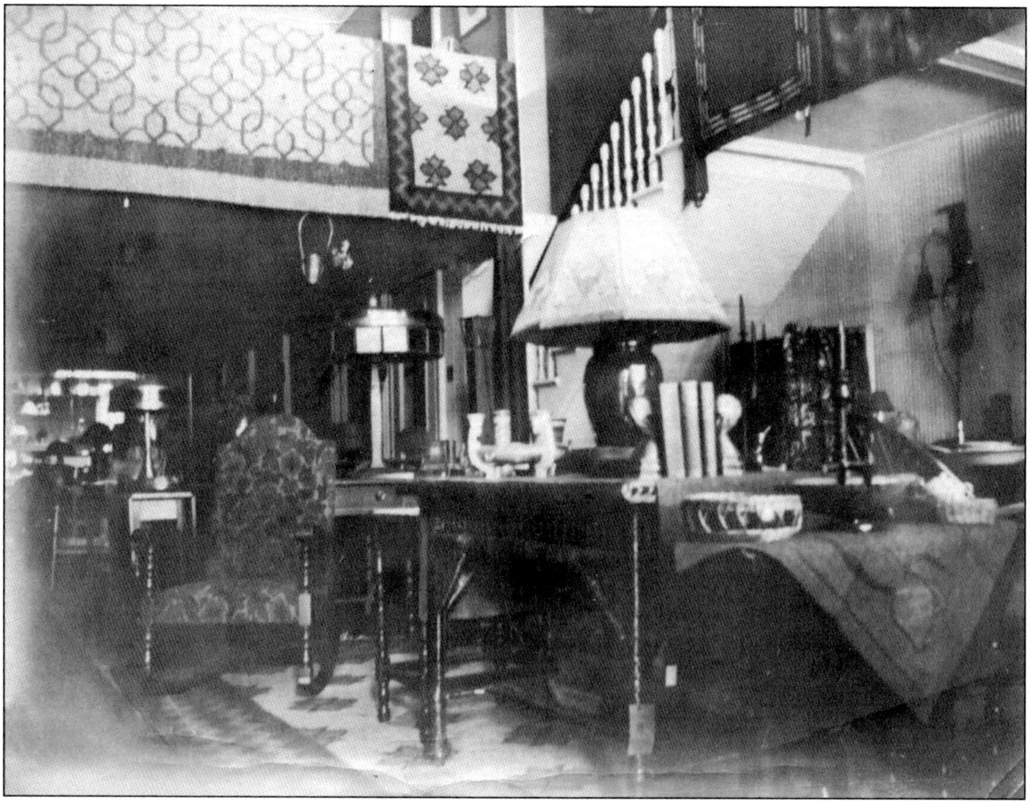

CRAFTSMAN ADVERTISING DEPARTMENT

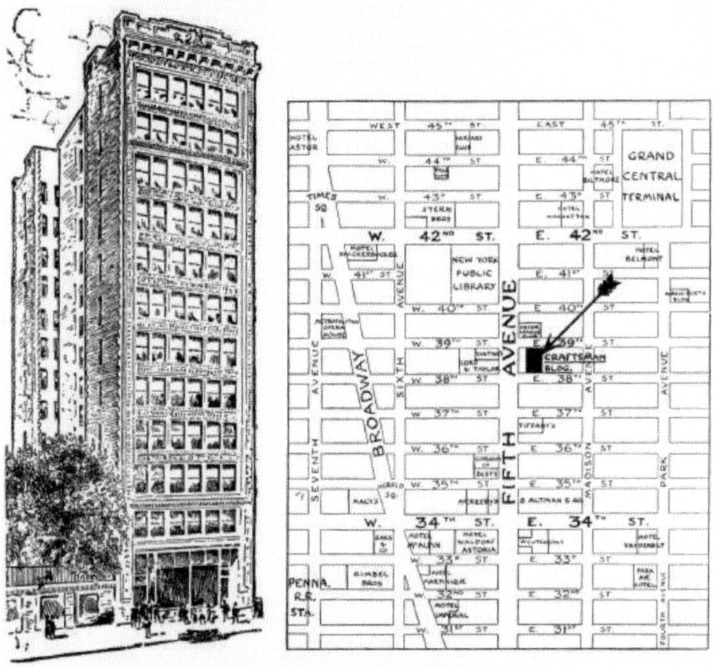


# When You Come To New York—

be sure to visit the Craftsman Building no matter how limited your time may be. Make it a point to come to the Building at the very first opportunity; if you put off your visit to the last minute, you are certain to regret it—there are so many things of interest here which you will wish you had more time to examine. The Craftsman Building is really a treasure-house for the homelover. The four floors devoted to furniture and furnishings are worth coming many miles to see, while for anyone about to build or remodel or redecorate, the exhibits in the Homebuilders' Exposition are of incalculable value as a means of saving time, money and trouble. Homebuilders who are wise in looking after their own best interests should plan to spend several days in studying the Exposition.

The Craftsman Building is conveniently situated in the best known, most accessible part of New York, within a few blocks of the Waldorf, Vanderbilt, Biltmore, Belmont, McAlpin, Knickerbocker, and other leading hotels, near the principal clubs and theatres, and in the midst of the famous Fifth Avenue shopping section which the visitor is sure to frequent. Craftsman readers are cordially urged to take advantage of this convenient location by making the Craftsman Building their headquarters while in New York, using the delightful club rooms and rest rooms freely, both for their own comfort and as a rendezvous for meeting friends, etc. The Craftsman Restaurant on the top floor, of which you have heard so much, will prove for you, as it has for so many others, one of the most delightful eating-places you have ever visited.

And by all means, when you do come, let us know that you are a Craftsman reader, and give us the opportunity to welcome you to your "own club-house."

Gustav Stickley

4 a

Always ambitious and striving to reach new heights, Stickley was naturally drawn to the bigger stage that New York City offered. In 1905, he began establishing a branch showroom on West Thirty-Fourth Street, making regular trips between Syracuse and Manhattan. By 1912, eager to make an even bigger mark on the world, he rented an entire 12-story building on East Thirty-Ninth Street, shown here in an advertisement from *The Craftsman* in April 1915. Furniture was artistically arranged in complete room settings, but the Craftsman Building was more than a mere showroom. Stickley sought to use the Craftsman Building to underscore his roles as tastemaker and social critic, in addition to that of successful businessman and manufacturer. Educational events were offered in a lecture hall in the building, and customers were also treated to crafts demonstrations. There was also a library in the building. A children's playroom on the 11th floor spoke to his recurring theme of family life, and a restaurant on the 12th floor was operated in keeping with the magazine's theme of wholesome living.

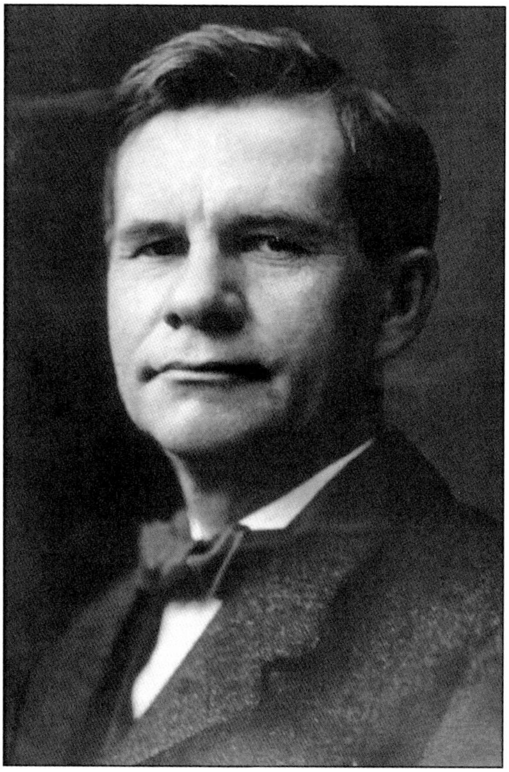

This ambitious man continued funding many simultaneous ventures. An undated photograph of Stickley's factory workers in Eastwood, New York, shows only a few of the people on his payroll. In 1908, while buying land for Craftsman Farms, Gustav opened another retail store in Boston. By the time of this 1910 portrait, he operated the Craftsman Company of New York City, the Craftsman Home Building Company, and the Craftsman Publishing Company. In 1911, his family settled into Craftsman Farms, and in 1912, he opened a retail branch in Washington, DC. In 1913, in a fatal overreach, he signed a 21-year lease on the Craftsman Building. The annual rent was approximately $60,000 at a time when an average annual salary was $564. (Below, courtesy the Stickley Museum, L. & J.G. Stickley, Inc.)

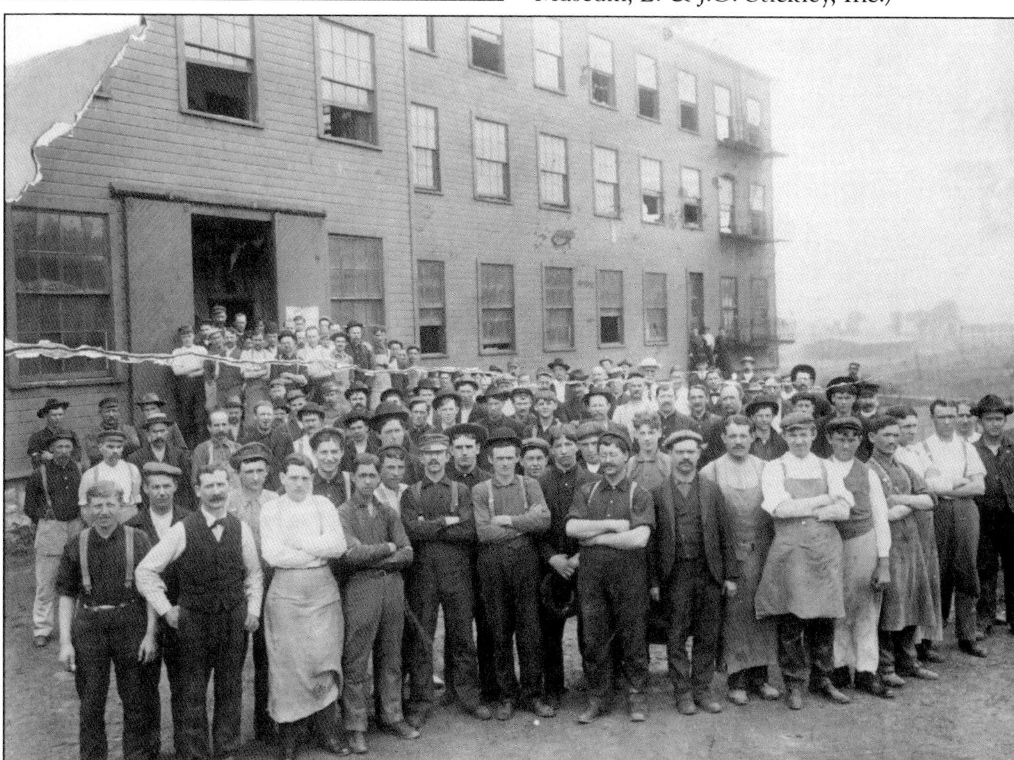

# *Two*

# Landscape and Buildings

In many ways, Gustav Stickley's creation of Craftsman Farms was the sum of a lifetime of beliefs and dreams. From his rugged early life in the Wisconsin frontier through his years in the business world, he never lost his love of nature. As a father, he found great pleasure in taking his children on strenuous hikes and even taught his daughters to fish. Craftsman publications were filled with articles on art, handcrafts, design, and the charm of country life but also thoughtful articles about education, social theories, and the importance of environmental preservation.

When Charles Wagner, author of the book *The Simple Life*, was invited to America by Pres. Theodore Roosevelt, Stickley made a point of inviting Wagner to present a lecture at the Craftsman Building. Wagner later wrote, "To build houses worthy to be the social centre of the family [and] to furnish them with objects at once useful, practical and capable of speaking to the heart—such is Mr. Stickley's high ideal."

In 1909, when work was underway at Craftsman Farms, Stickley published the book *Craftsman Homes*. It included an article entitled "The Craftsman Idea," which stated, "while we take the greatest interest in all efforts toward reform in any direction, we remain firm in the conviction that the root of all reform lies in the individual and that the life of the individual is shaped mainly by home surroundings and influences and by the kind of education that goes to make real men and women instead of grist for the commercial mill."

Of course, Stickley himself had been part of that commercial mill in what could be seen as excessive or ruthless ambition. But whatever Stickley's failings might have been in striving for commercial success, these are far outweighed by his overarching desire to be moral, ethical, and just. The very existence of an idealistic and peaceful place like Craftsman Farms stands as quite a testimony to the Thoreau-like values that were central to his Craftsman ethic.

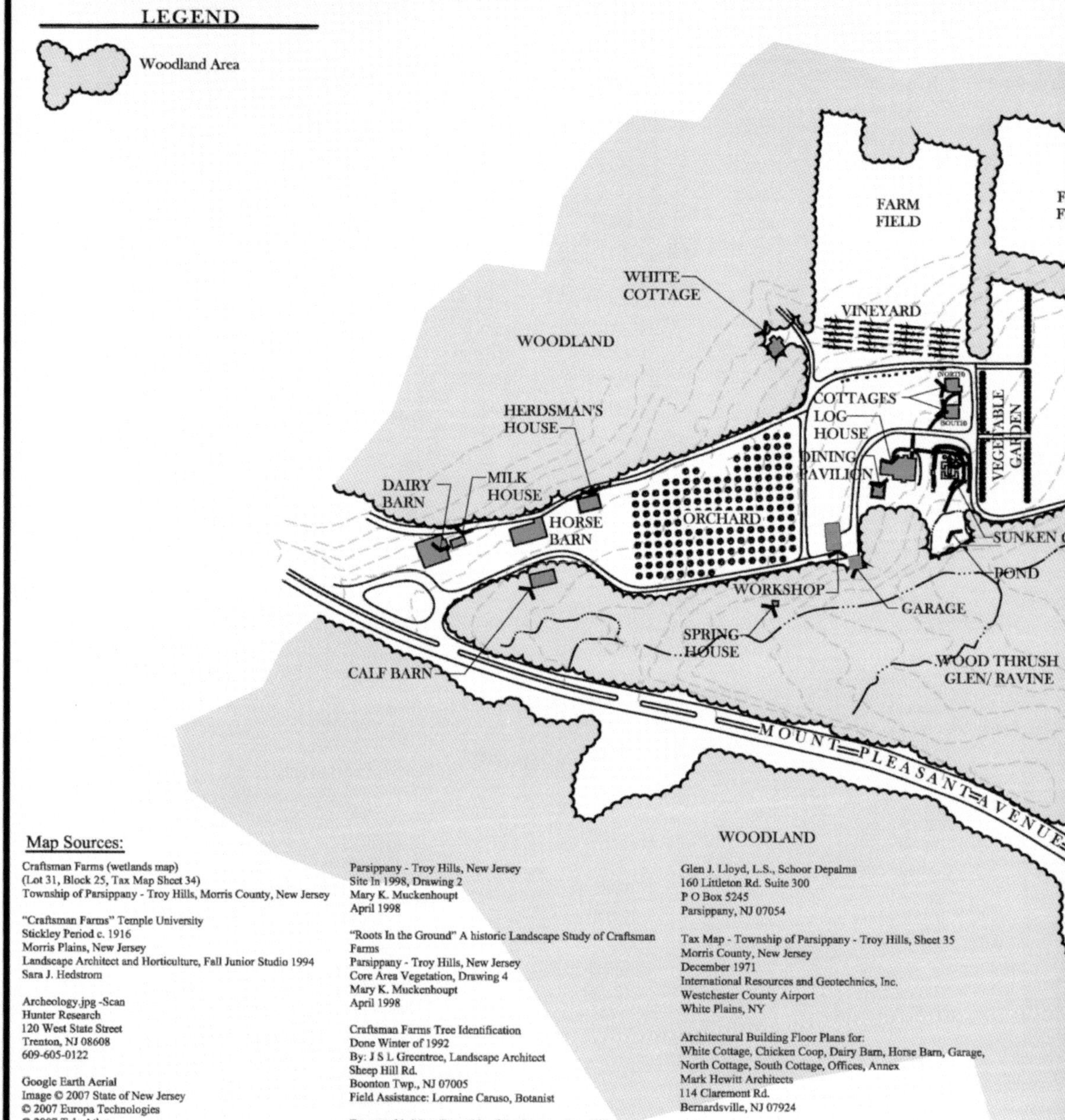


## Map Sources:

Craftsman Farms (wetlands map)
(Lot 31, Block 25, Tax Map Sheet 34)
Township of Parsippany - Troy Hills, Morris County, New Jersey

"Craftsman Farms" Temple University
Stickley Period c. 1916
Morris Plains, New Jersey
Landscape Architect and Horticulture, Fall Junior Studio 1994
Sara J. Hedstrom

Archeology.jpg -Scan
Hunter Research
120 West State Street
Trenton, NJ 08608
609-605-0122

Google Earth Aerial
Image © 2007 State of New Jersey
© 2007 Europa Technologies
© 2007 Tele Atlas

Craftsman Farms Existing Site Plan
March 6, 1992
School of Architecture, New Jersey Institute of Technology
Advanced Options Studio, 1992

Craftsman Farms - First Floor Plan (Log House)
May 9, 1992
School of Architecture, New Jersey Institute of Technology
Advanced Options Studio, 1992

"Roots In the Ground" A historic Landscape Study of Craftsman Farms
Parsippany - Troy Hills, New Jersey
Site In 1998, Drawing 2
Mary K. Muckenhoupt
April 1998

"Roots In the Ground" A historic Landscape Study of Craftsman Farms
Parsippany - Troy Hills, New Jersey
Core Area Vegetation, Drawing 4
Mary K. Muckenhoupt
April 1998

Craftsman Farms Tree Identification
Done Winter of 1992
By: J S L Greentree, Landscape Architect
Sheep Hill Rd.
Boonton Twp., NJ 07005
Field Assistance: Lorraine Caruso, Botanist

Topographic Map, Township of Parsippany - Troy Hills
Morris County, New Jersey
Compiled By: International Resources and Geotechnics, Inc.
Contour Internal = 2', December 1972
Sheet No. 3652

Topographic Map, Township of Parsippany - Troy Hills
Morris County, New Jersey
Compiled By: International Resources and Geotechnics, Inc.
Contour Internal = 2', March 1975
Sheet No. 3649
Tax Map, Township of Parsippany - Troy Hills - Sheet 34
Morris County, New Jersey, October 2001

Glen J. Lloyd, L.S., Schoor Depalma
160 Littleton Rd. Suite 300
P O Box 5245
Parsippany, NJ 07054

Tax Map - Township of Parsippany - Troy Hills, Sheet 35
Morris County, New Jersey
December 1971
International Resources and Geotechnics, Inc.
Westchester County Airport
White Plains, NY

Architectural Building Floor Plans for:
White Cottage, Chicken Coop, Dairy Barn, Horse Barn, Garage, North Cottage, South Cottage, Offices, Annex
Mark Hewitt Architects
114 Claremont Rd.
Bernardsville, NJ 07924

Craftsman Farms 1938 Aerial - No Source Information

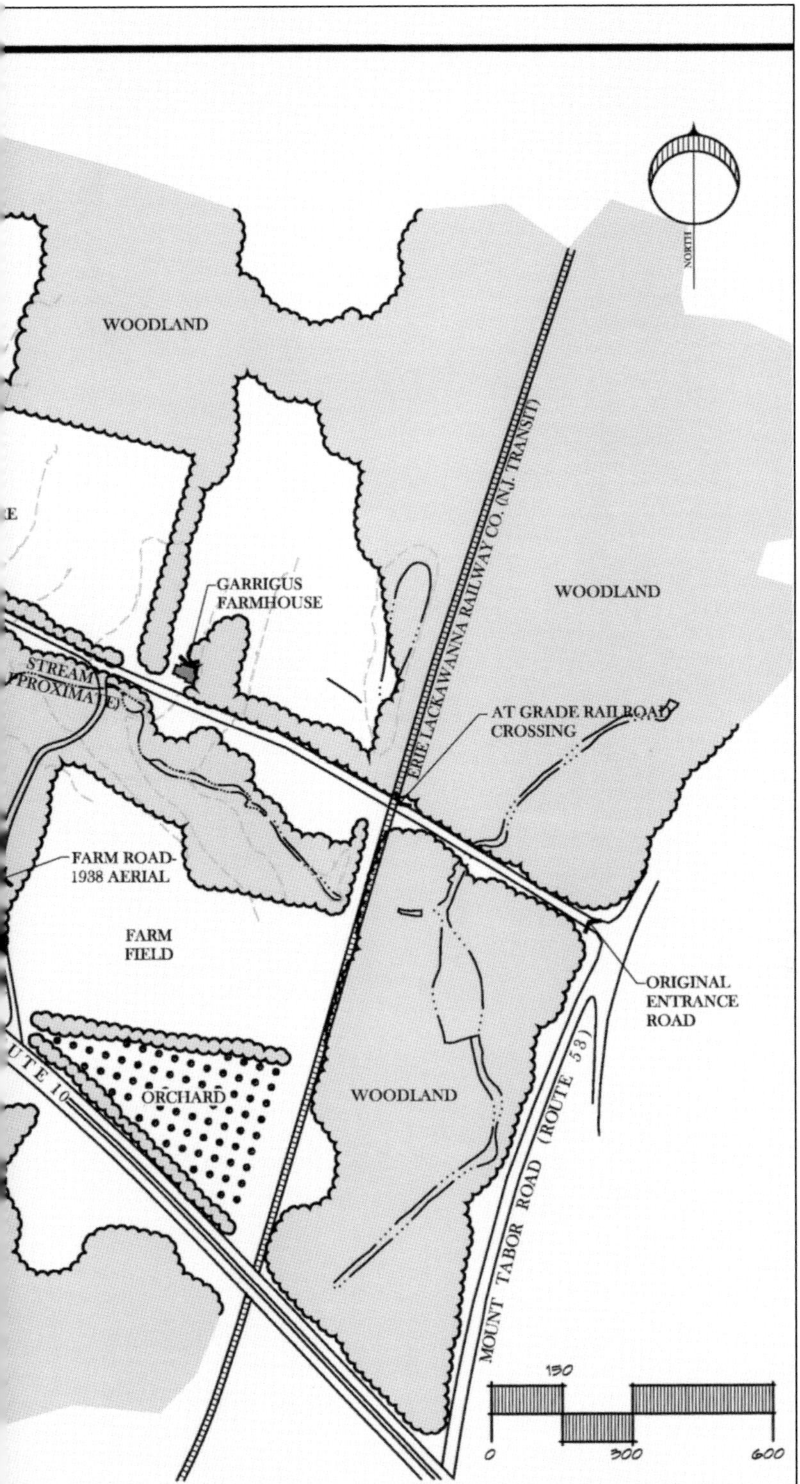


This map shows the core of the 650 acres of Craftsman Farms. The estate was made up of various tracts of land that Stickley purchased over a period of time beginning in 1908. These parcels included farmland, woodland, and wetland. He felt it was important to keep much of the property in its natural state and left the wetland area, known as Wood Thrush Glen, virtually untouched. Much of the woodlands too were allowed to grow freely and maintain natural habitats for wildlife. The farmland and farm buildings were carefully situated in relation to the natural curve of the gently rolling hills so that they nestled comfortably and naturally within the landscape. (Courtesy B.W. Bosenberg & Co., Inc.)

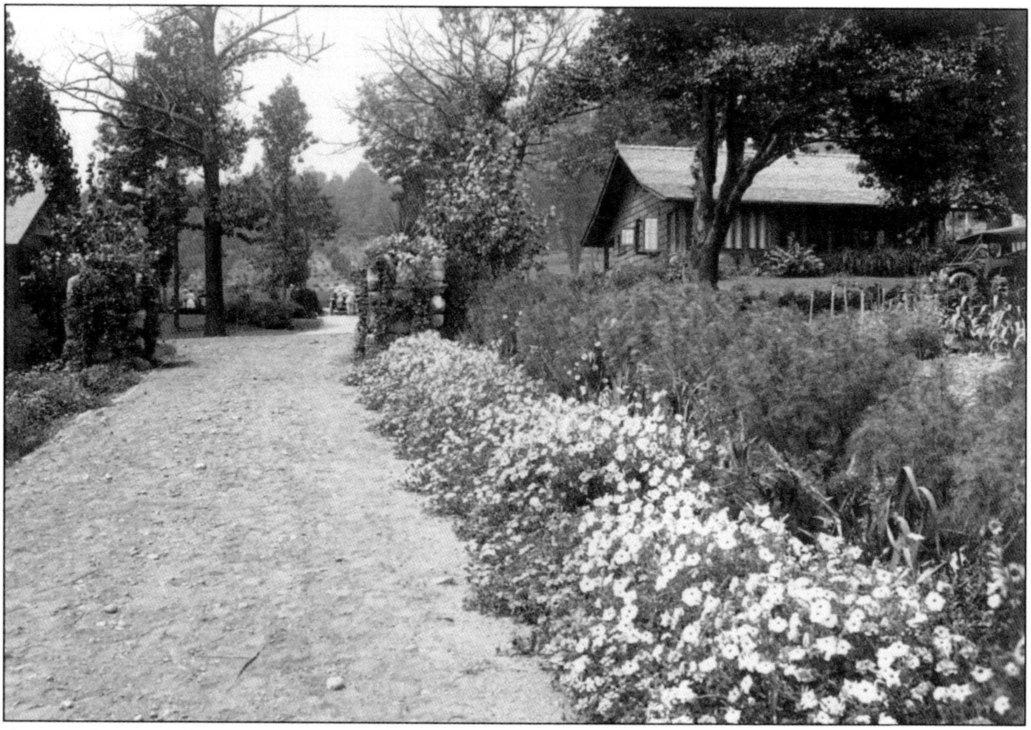

An article entitled "Craftsman Farms: Its Development and Future" in the October 1913 *Craftsman* noted that the flower beds and lawns near the houses at Craftsman Farms gave "contrast and meaning to the whole property. . . . [Such] contrasts are essential to success." That message is underscored by the profusion of lush plantings in this c. 1917 photograph of the road near the cottages.

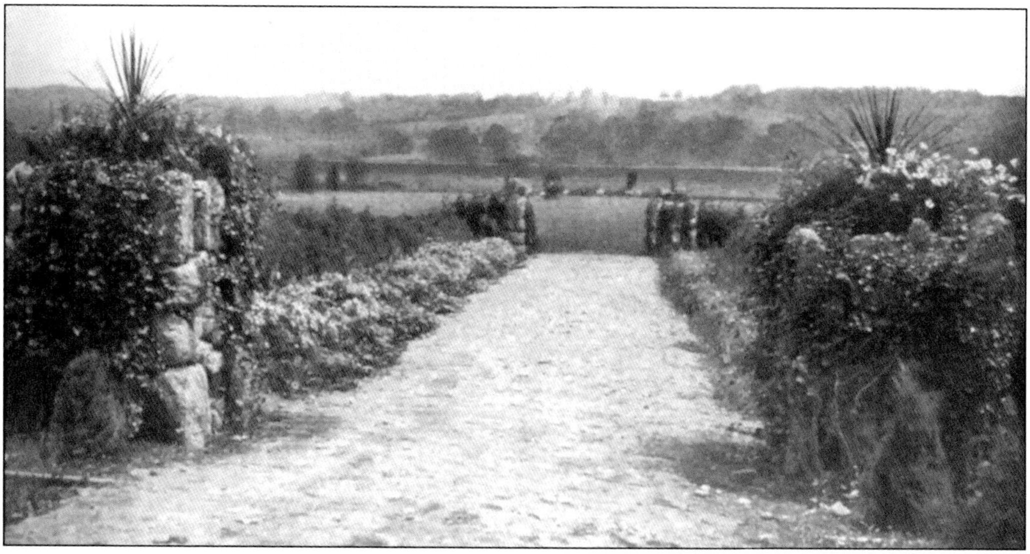

Located at the end of the road that passes the cottages, this pathway is marked by two stone piers. English ivy spills out over the stones and small densely packed blossoms mark the pathway out to a second pair of piers in the distance in this 1913 photograph. The pathway can be seen in the map on the previous page as it bisects the vegetable garden.

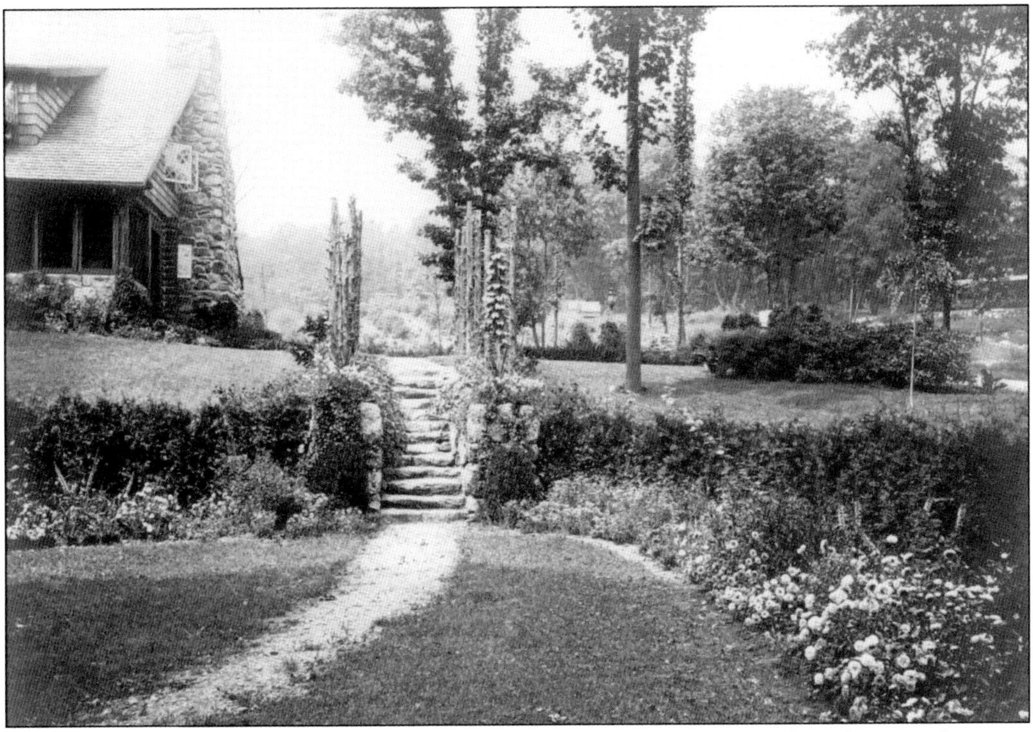

This 1913 photograph was taken from the rose garden, which was situated in a large sunken area below the front of the log house. A curving fieldstone retaining wall is barely visible behind the plantings that edge the garden, and rustic stone steps lead up toward the log house, which is partially visible on the left.

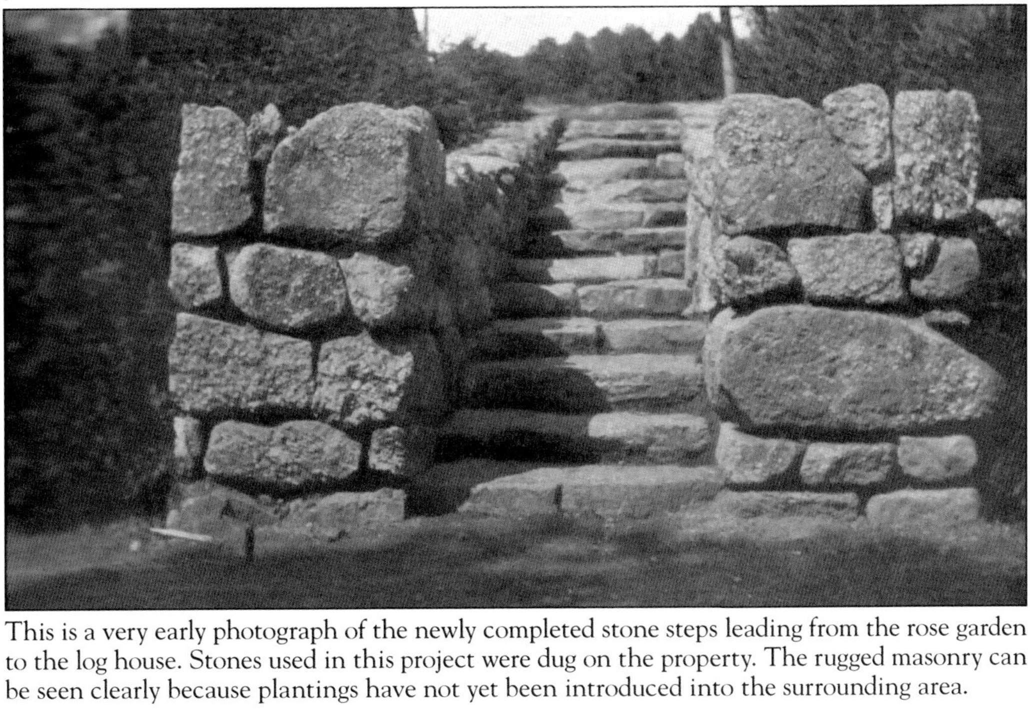

This is a very early photograph of the newly completed stone steps leading from the rose garden to the log house. Stones used in this project were dug on the property. The rugged masonry can be seen clearly because plantings have not yet been introduced into the surrounding area.

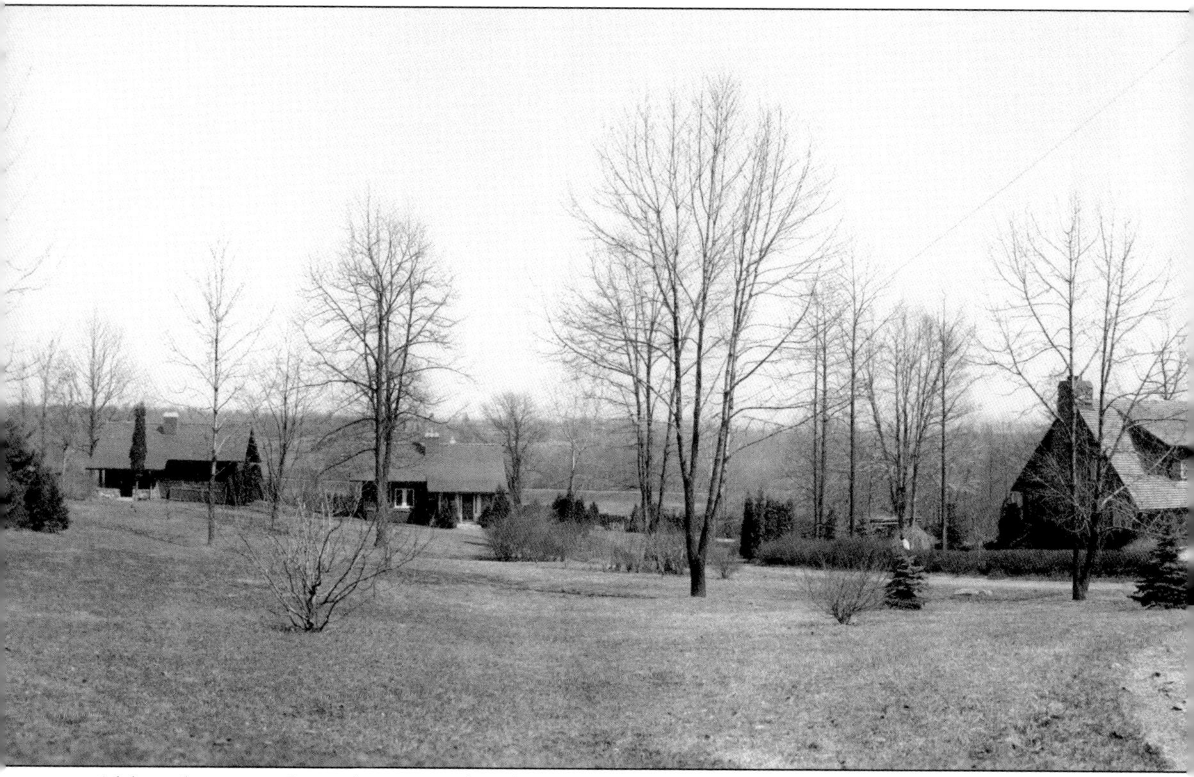

Although it was taken a few years after the Stickley family left Craftsman Farms, this photograph is the only existing image of the central area of Craftsman Farms. Taken from a hillside looking to the east, this image is especially useful because the bare trees allow a clearer view of the way Stickley grouped the houses together as a family compound. He was quoted as saying, "I like the houses to be grouped together. It makes life simpler, more friendly." On the left of the panorama,

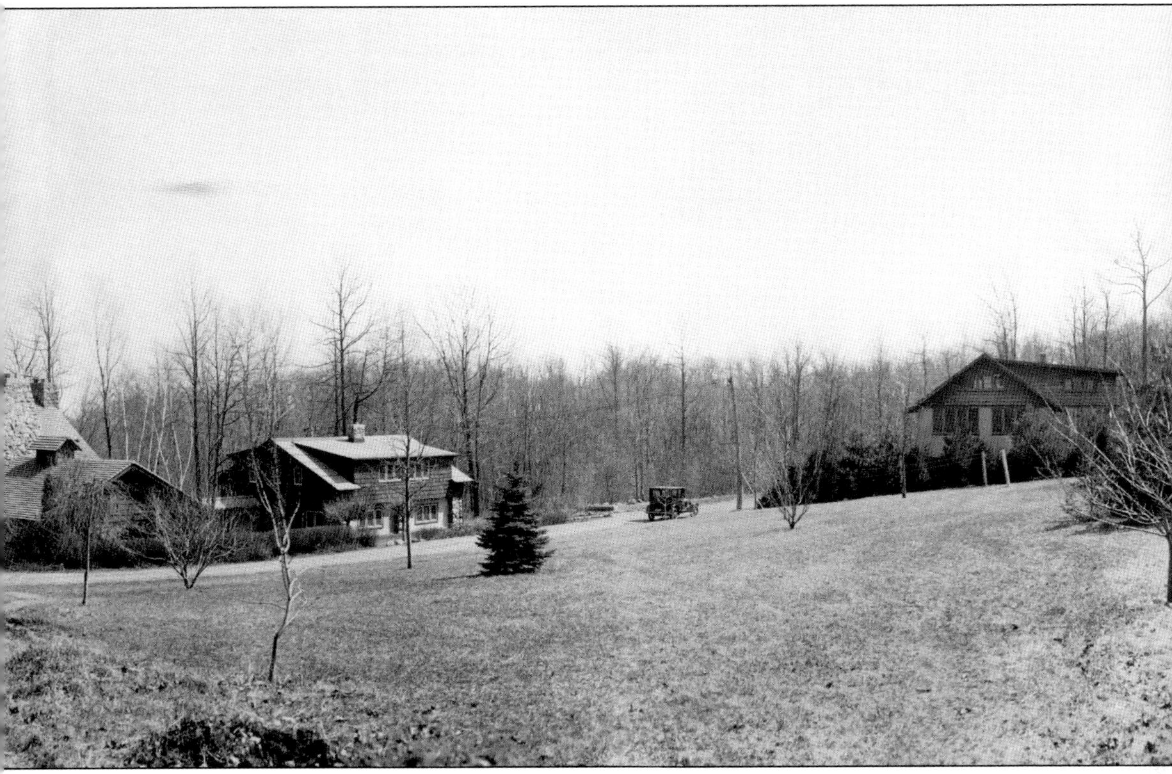

two cottages are shown across the road from the log house, which is partially visible on the same page. This image shows the kitchen wing of the log house, which is connected to a newer building known as the annex. In Stickley's time, the annex area had been an outdoor dining pergola. It was later enclosed, leaving the pergola's stone piers intact as part of the new building. A motorcar can be seen approaching the garage and workshop on the far right of the photograph.

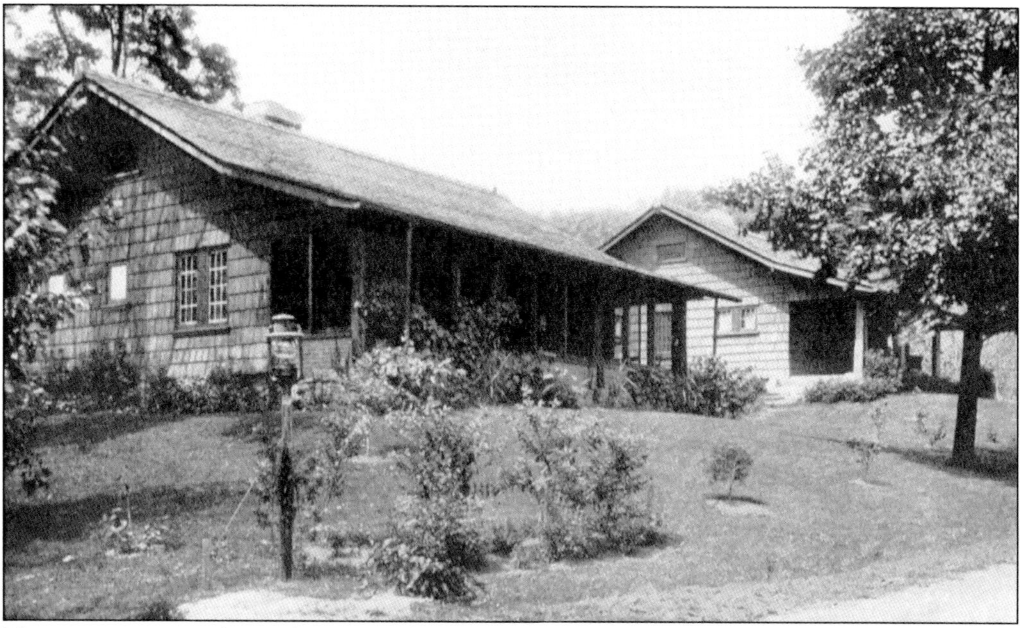

The twin cottages shown here were the first to be built at Craftsman Farms and were where the Stickley family initially lived when they moved from Syracuse. At various times after their weddings, daughters Barbara, Mildred, and Marion each lived in one of the cottages with their new husbands. The exterior walls of the cottages were covered in brown cypress shingles, which blended harmoniously with the outdoor landscape. In what Stickley described as "a cheery note of color," the cottage roofs were made of rough red slate topped with red terra-cotta roof ridge tiles. The porches also had a touch of color in their red-tinted poured cement floors. Each cottage had a large copper-hooded fireplace and warm brown chestnut paneling inside.

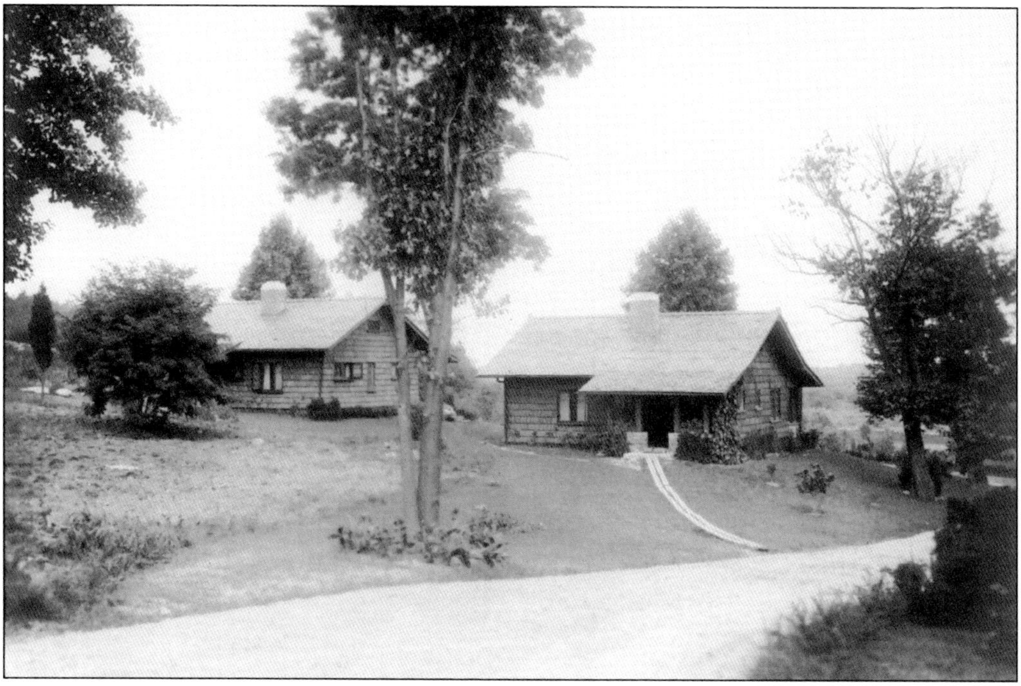

This easterly view shows the landscaped lawn of the family compound, with the entrance to the log house on the right and a portion of one of the twin cottages on the left. Continuing down the hill to the east were the vegetable gardens. Just out of view to the north of the cottages was the vineyard.

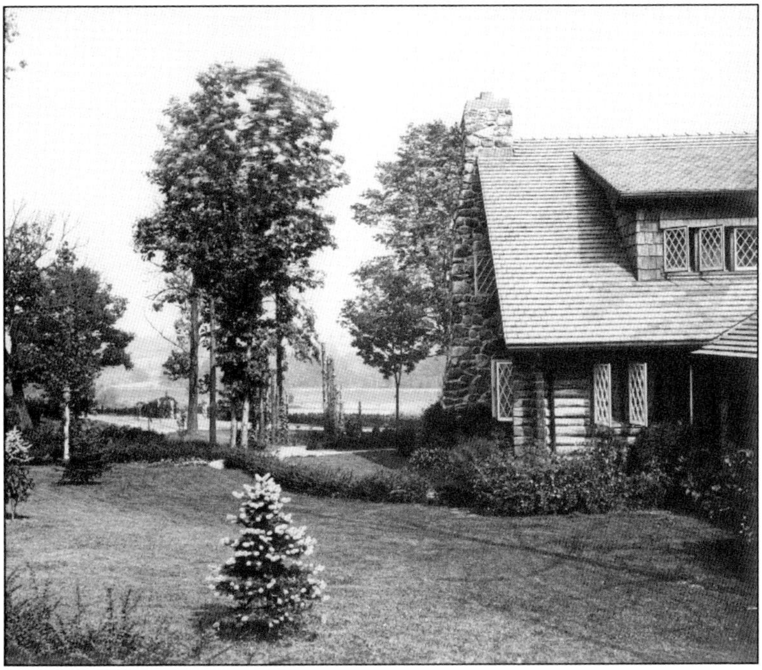

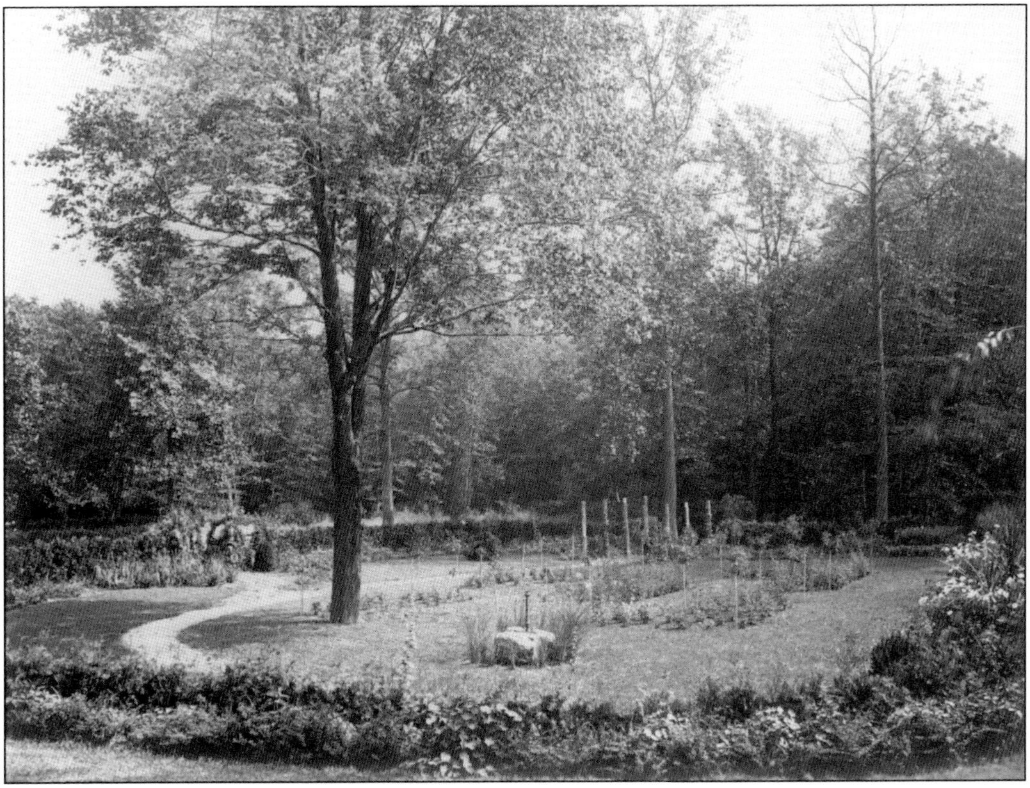

A view of the rose garden itself shows an abundance of other well-cared-for plants in addition to the roses, which are planted in formal flower beds. The more formal landscaping and profusion of flowers make a lovely contrast with the wooded area to the right of the picture.

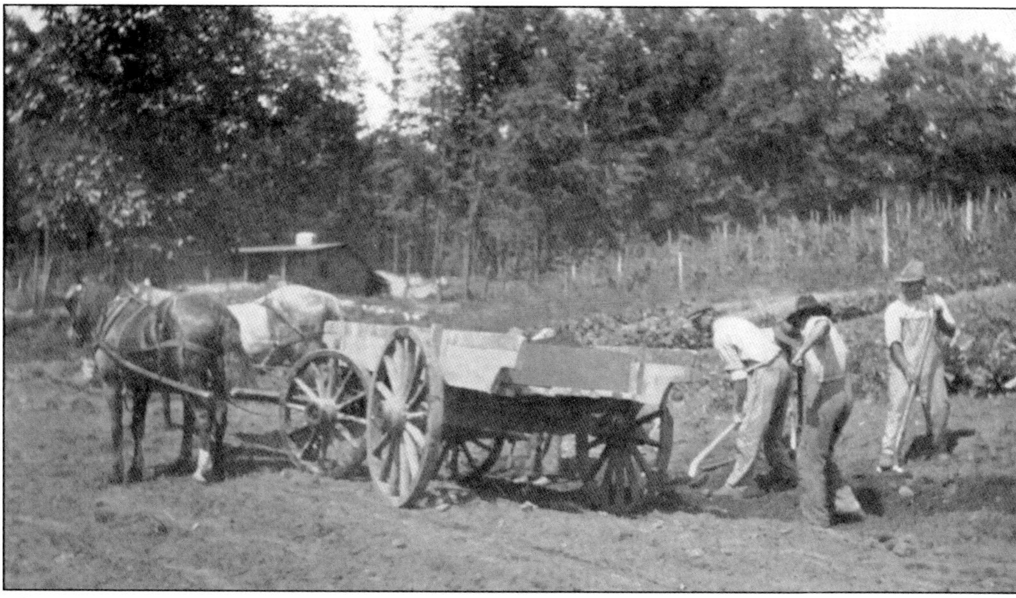

This 1912 photograph shows workers clearing the ground for a road at the edge of the family compound. The work cart may have been used to haul away the larger boulders that would be used in masonry on the property. The vineyard can be seen on the hillside behind the men, and the white cottage can be seen behind the horses.

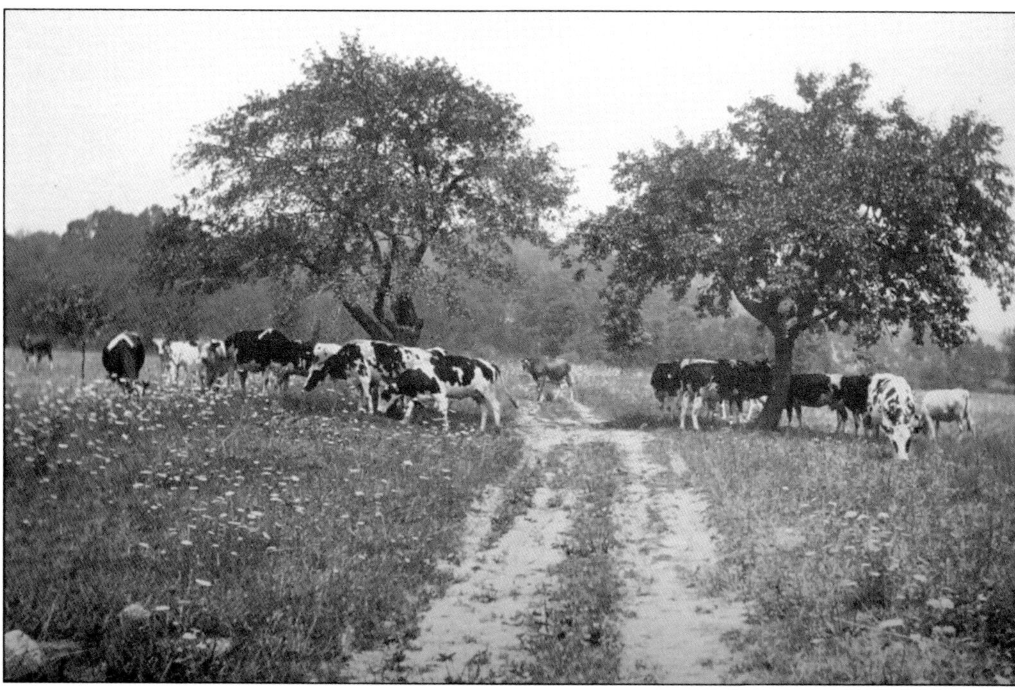

In this 1913 photograph, Stickley's herd of Holstein cows can be seen grazing freely in the fields to the east of the vegetable gardens. Stickley noted that he chose Holsteins because the breed, which originated in Holland, were, "larger, stronger, healthier, and longer-lived than any other cow on earth."

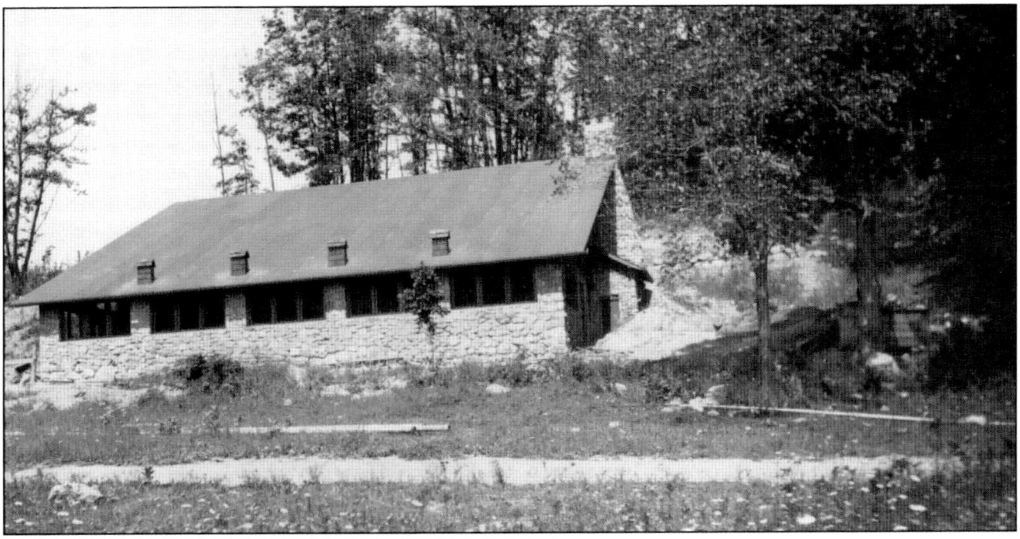

Stickley established a grouping of farm buildings to the west of the family compound. The westernmost building was the stone cow barn, shown here, which was designed with both the lay of the land and the health of the cows in mind. Heat from a Craftsman fireplace helped protect the herd in the winter. The stone chimney appears on the right of the barn.

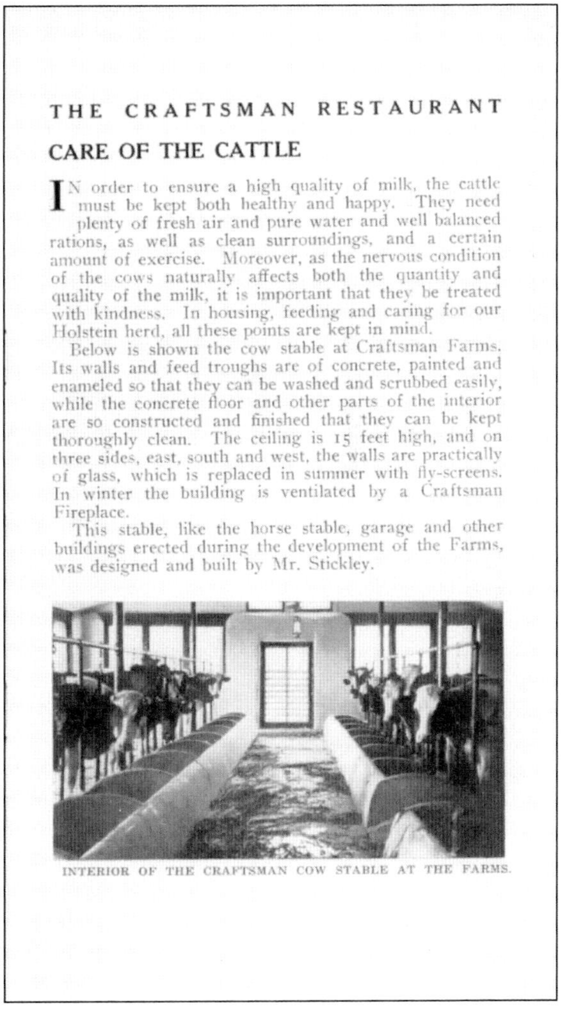

THE CRAFTSMAN RESTAURANT

CARE OF THE CATTLE

IN order to ensure a high quality of milk, the cattle must be kept both healthy and happy. They need plenty of fresh air and pure water and well balanced rations, as well as clean surroundings, and a certain amount of exercise. Moreover, as the nervous condition of the cows naturally affects both the quantity and quality of the milk, it is important that they be treated with kindness. In housing, feeding and caring for our Holstein herd, all these points are kept in mind.

Below is shown the cow stable at Craftsman Farms. Its walls and feed troughs are of concrete, painted and enameled so that they can be washed and scrubbed easily, while the concrete floor and other parts of the interior are so constructed and finished that they can be kept thoroughly clean. The ceiling is 15 feet high, and on three sides, east, south and west, the walls are practically of glass, which is replaced in summer with fly-screens. In winter the building is ventilated by a Craftsman Fireplace.

This stable, like the horse stable, garage and other buildings erected during the development of the Farms, was designed and built by Mr. Stickley.

INTERIOR OF THE CRAFTSMAN COW STABLE AT THE FARMS.

In describing the construction of his modern farm buildings, Stickley said, "We are constructing these and other buildings so that work will be simplified to the utmost degree." The feed troughs in the cow barn, which are shown here, were coated with white enamel for easier cleaning. An early electric milking system was also used to lighten the work of farmhands.

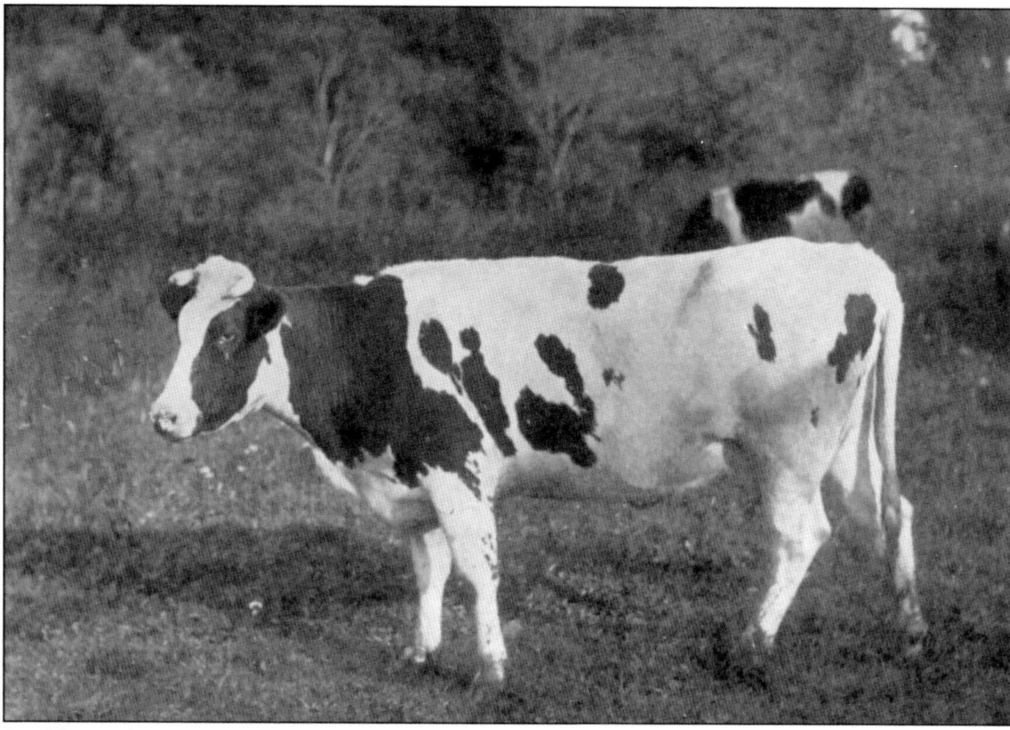

Stickley took enormous pride in his thoroughbred registered Holstein cows. He even published this picture of one of his favorites. She was named Woodcrest Rachel and apparently was so healthy that she produced 19,443 pounds of milk and 851 pounds of butter in a single year.

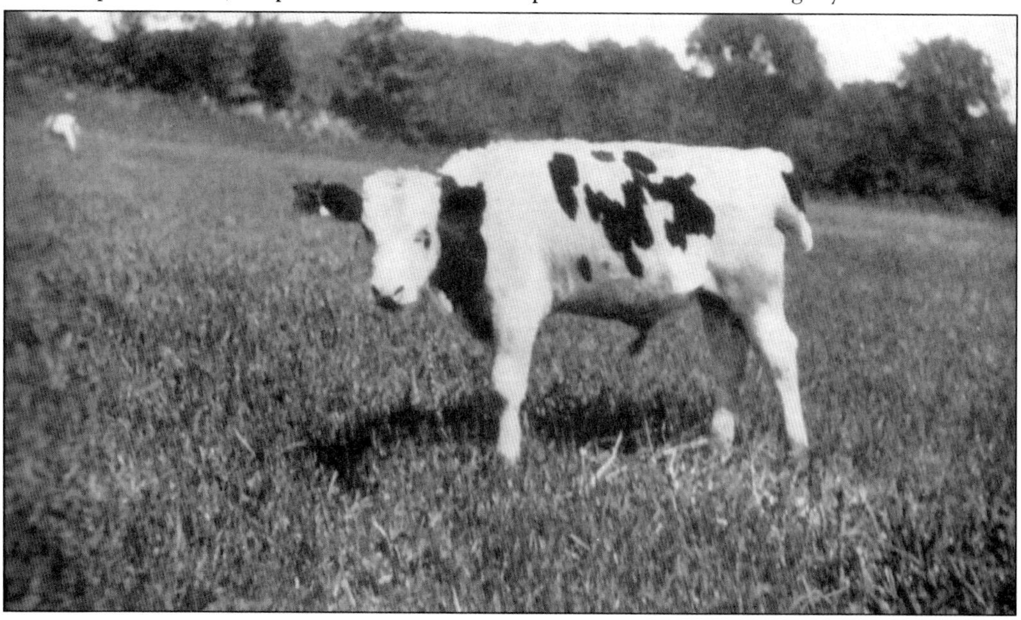

Stickley also published this charming photograph of the first calf to be born at Craftsman Farms. The calf was born on February 22 and named George Washington because it was Washington's birthday. There was a separate calf barn on the property, but there are no known photographs of the building during Stickley's time.

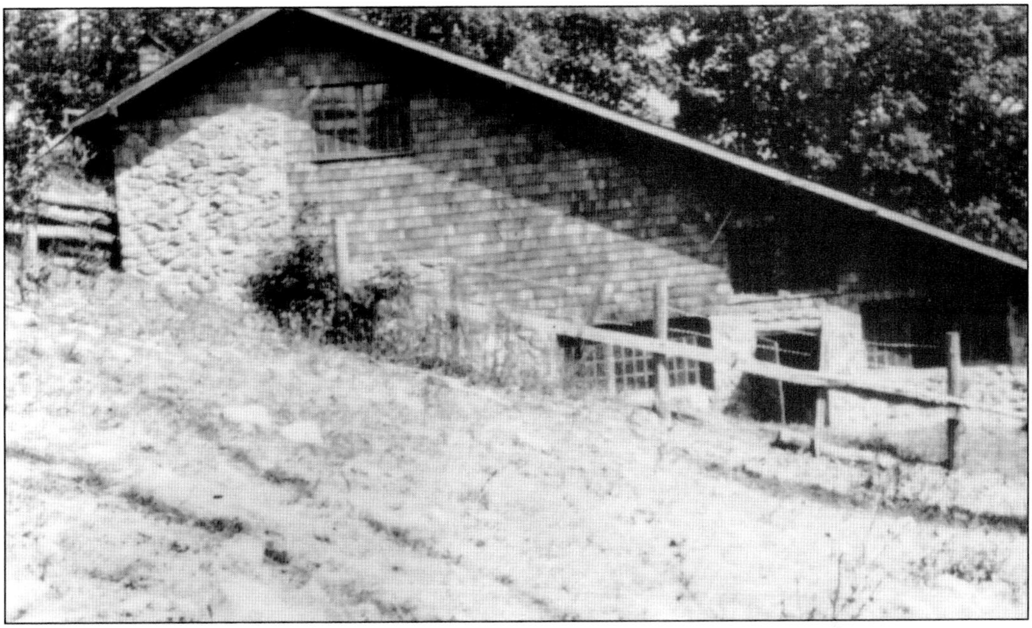

This image of the side of the barn is from the 1920s, shortly after the property was sold. It shows the barn neatly nestled into the existing hillside. Paths connected the barn to the milk house and stables on the front, lower side. A higher path in the back allowed the farmhands easy access to the barn's second-floor hayloft.

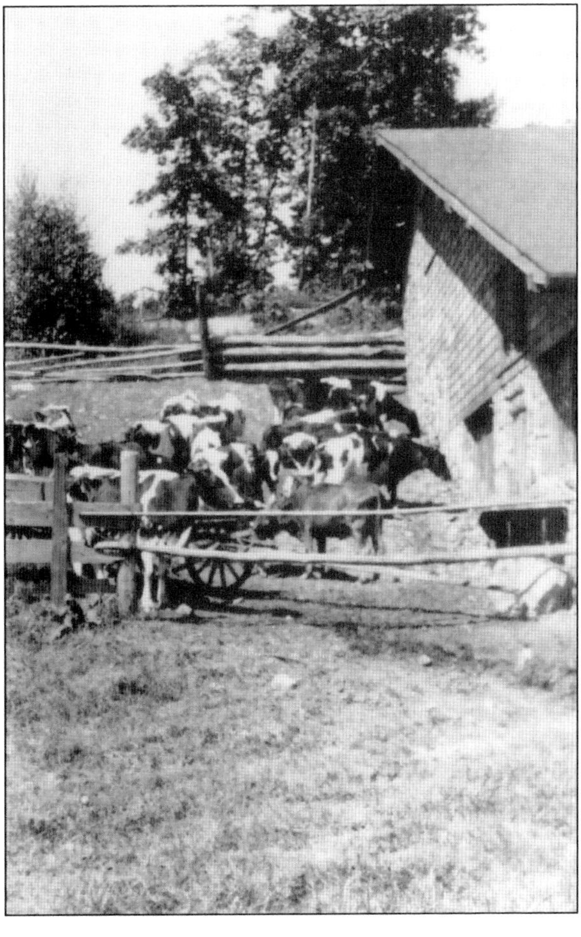

Also dating from the 1920s, this image shows the western side of the barn with the herd of cows gathered in, possibly for milking. Back at the milk house, their milk would have been cooled in metal milk cans placed into cold spring water and then bottled into pint- and quart-sized classic milk bottles bearing "Craftsman Dairy" in raised letters.

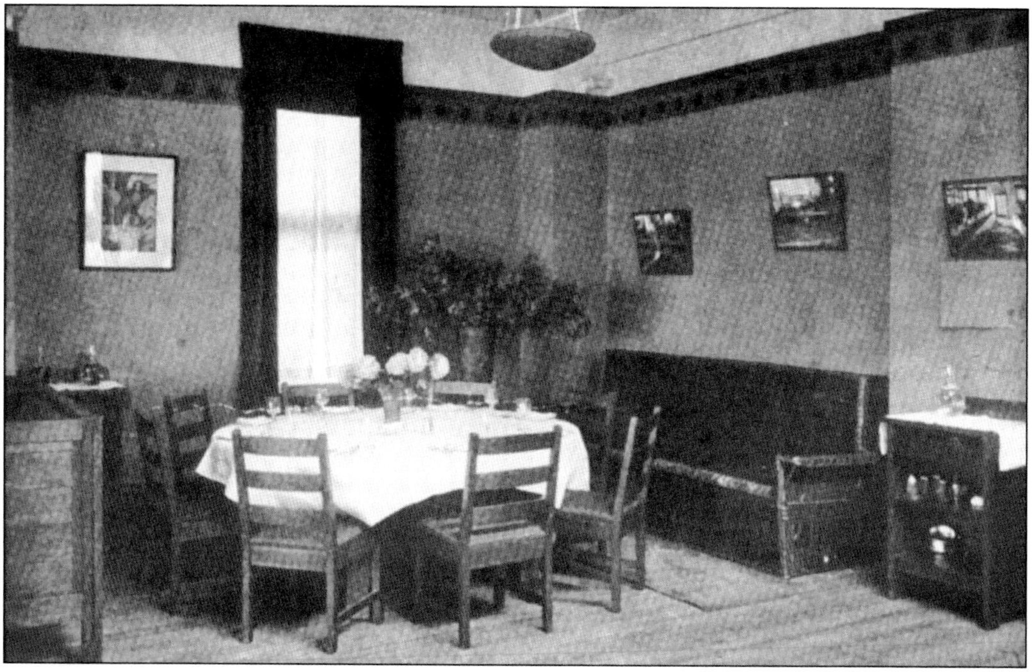

In talking about his Manhattan restaurant, Stickley said, "The restaurant will be intimately associated with Craftsman Farms, naturally, and all the wholesome products people crave in a city will be brought in every morning from the Farms." Close examination of this January 1914 magazine photograph reveals the picture of his cows in the barn is proudly displayed on the restaurant wall.

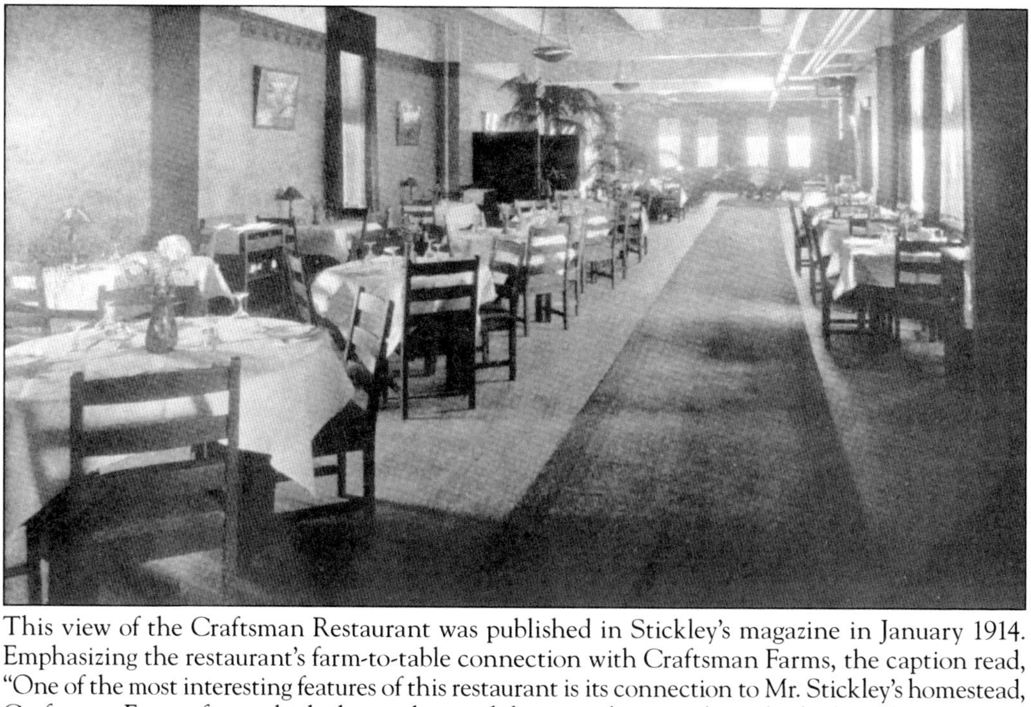

This view of the Craftsman Restaurant was published in Stickley's magazine in January 1914. Emphasizing the restaurant's farm-to-table connection with Craftsman Farms, the caption read, "One of the most interesting features of this restaurant is its connection to Mr. Stickley's homestead, Craftsman Farms, from which the poultry and dairy products are brought fresh each morning."

# *Three*

# THE LOG HOUSE

The centerpiece of Craftsman Farms is undoubtedly the massive log house, which is now widely considered to be one of the most significant buildings in the American Arts and Crafts movement. Despite the grand scale of the house—the living room and dining room are each 50 feet across—it does not exude any showiness or grandeur. In fact, the most common reaction visitors have to the home is that of peacefulness.

The house was originally conceived as the clubhouse for Stickley's planned school for boys. Fueled by the ideas of American educator John Dewey, Russian thinker Prince Kropotkin, and English artist John Ruskin, Stickley was inspired to create a progressive school at Craftsman Farms. He announced his plans for the school in the October 1908 issue of *The Craftsman*. Numerous articles about educational theories appeared in his magazine, including several by progressive educator Raymond Riordin.

The school would blend book learning with experiential learning. Students would study music, literature, and art, but would also experience manual labor making roads and building their own houses under the guidance of experienced carpenters and stonemasons.

Stickley's school would help build self-reliance and rugged individualism, which were highly regarded traits in Theodore Roosevelt's America. This was a time when the exhilarating imagery of the American West painted by Frederic Remington and Albert Bierstadt helped to instill such ideas deeply into the American consciousness. Stickley's school plans must have been impacted by all of this as well as memories of his own hardscrabble upbringing in the American frontier.

Sadly, after several attempts, he was unable to attract students to his school. He reworked the design of the clubhouse into a design for his family's home. Its form as a log house stands as a powerful reminder of the "land of promise" that Stickley had experienced in the American frontier. Though his original plan for a school never materialized, his home stands as a shining example of a way of life, and it can still teach valuable lessons.

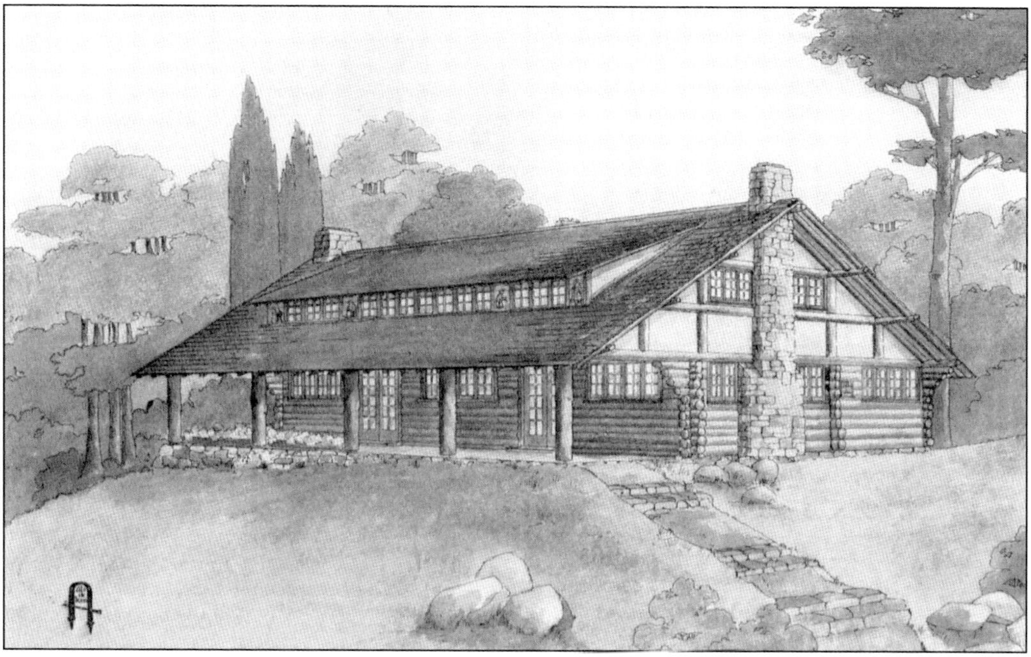

This drawing from *The Craftsman*, December 1908, was the original concept for the log house. At that time, Stickley anticipated building the log house as the clubhouse for his planned School for Citizenship. Some time after the construction started, he realized the school was not materializing as he had hoped. He then reworked the clubhouse design to serve as the family's home.

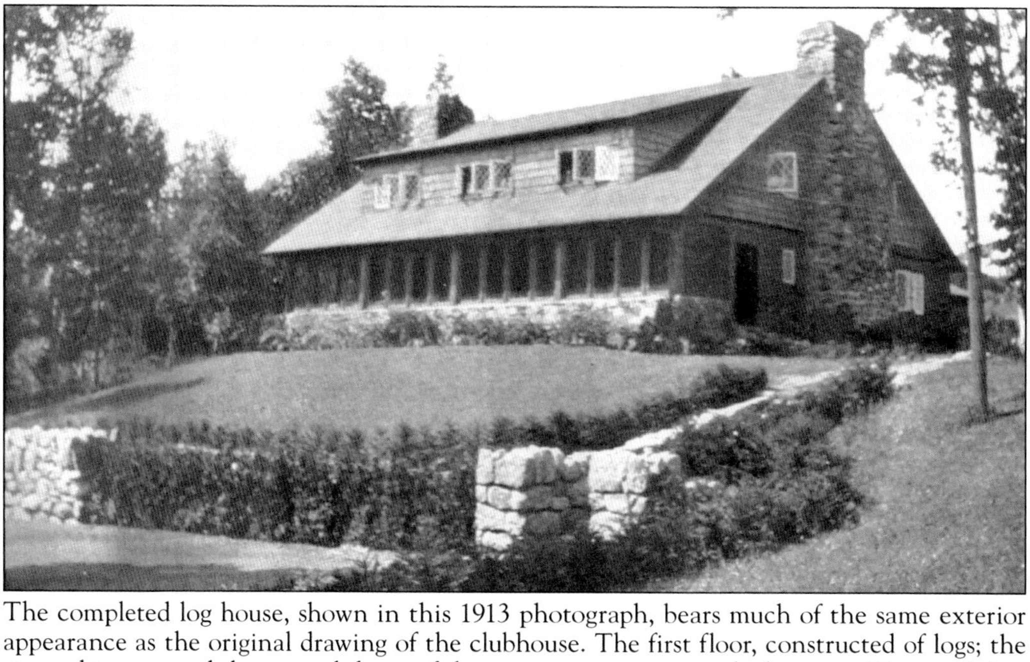

The completed log house, shown in this 1913 photograph, bears much of the same exterior appearance as the original drawing of the clubhouse. The first floor, constructed of logs; the stone chimney; and the general shape of the structure are very much the same. The porch has been enclosed, and the windows in the dormer have been changed.

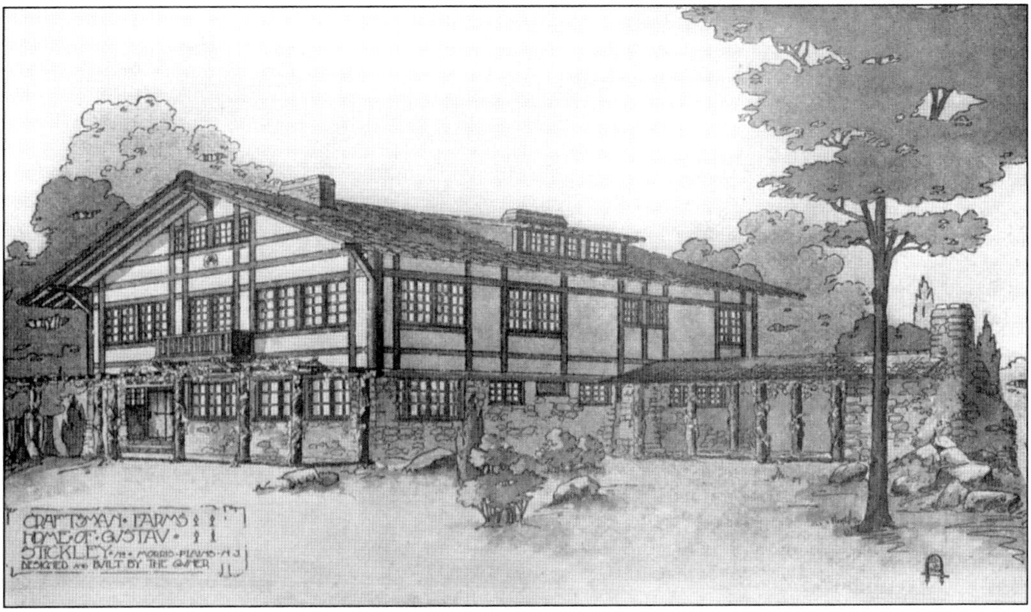


This drawing of the family home Stickley planned to build at Craftsman Farms is marked "Craftsman Farms, home of Gustav Stickley . . . designed and built by the owner." A dining pergola is shown attached to the rear of the house. Financial realities forced him to abandon the concept of building this home, but he incorporated some of the ideas into the redesigned log house.

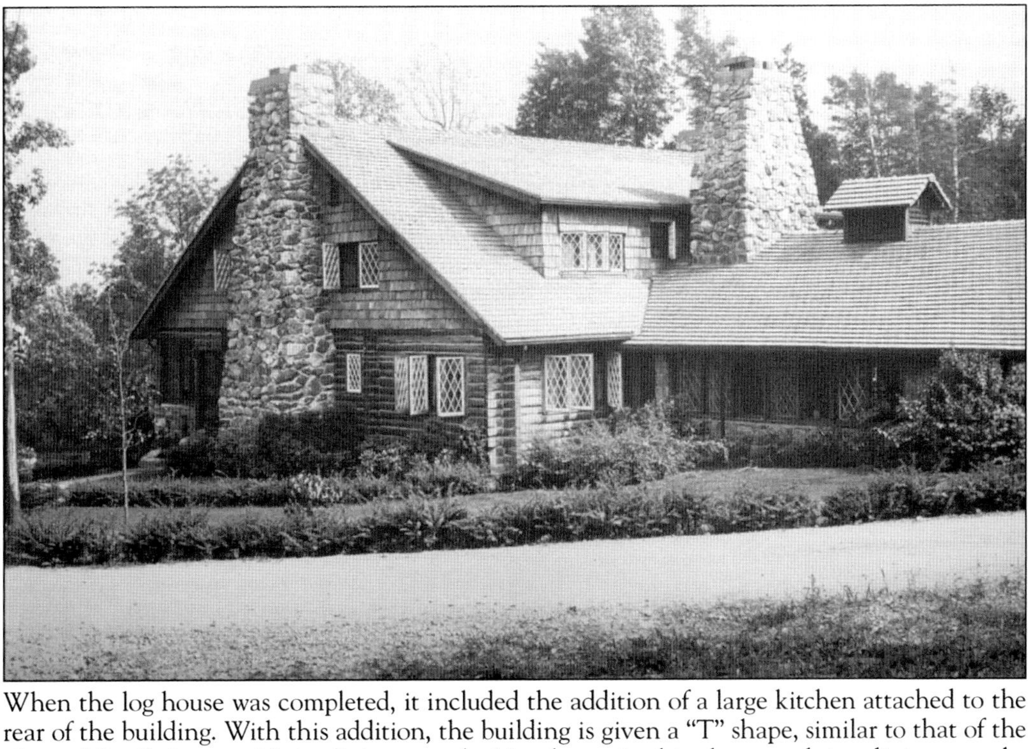

When the log house was completed, it included the addition of a large kitchen attached to the rear of the building. With this addition, the building is given a "T" shape, similar to that of the planned family home with its dining pergola. Not shown in this photograph is a dining pergola, which extended beyond the kitchen.

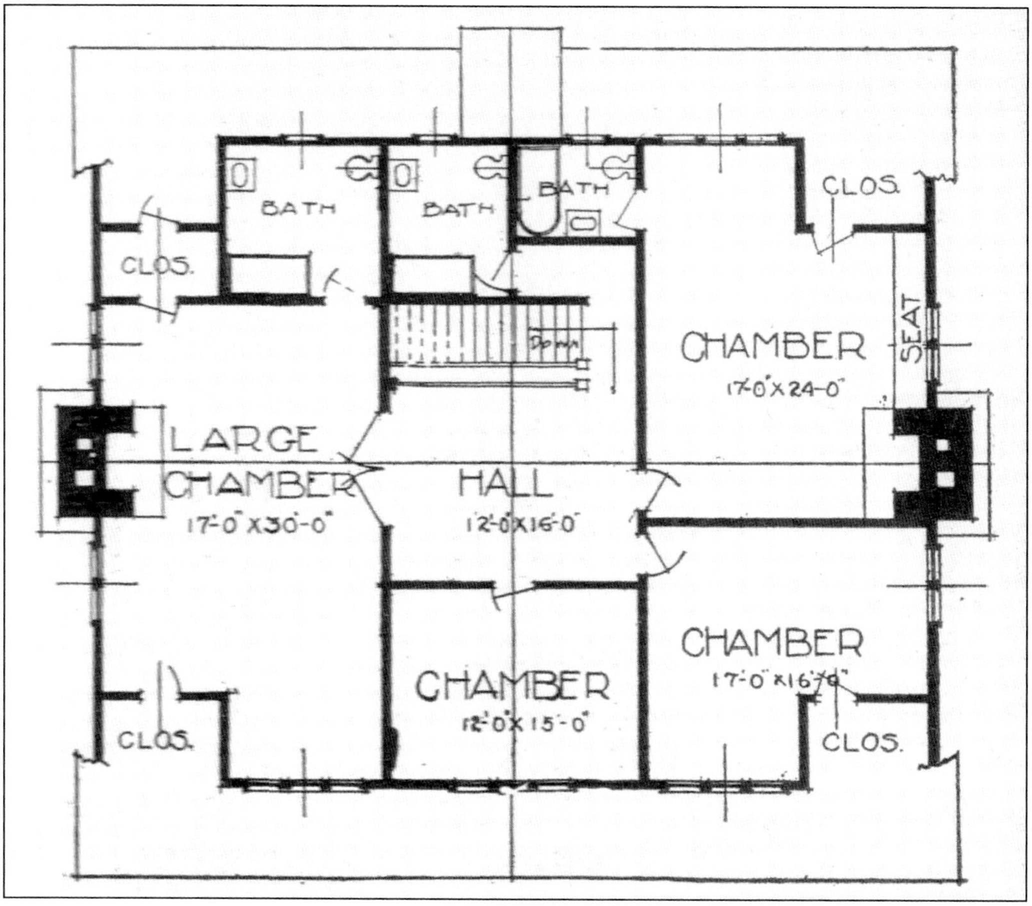


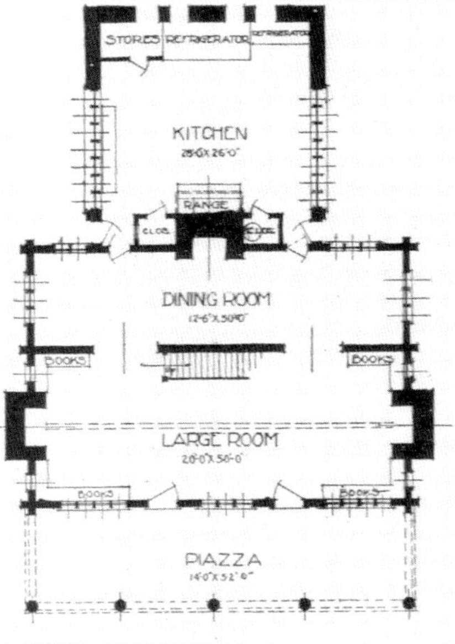


LOG HOUSE AT CRAFTSMAN FARMS: FIRST FLOOR PLAN

These hand-drawn floor plans show the final version of the log house. The smaller drawing shows the porch, identified here as a piazza, with two doors leading into the living room, which spans a full 50 feet across. The living room's two massive stone fireplaces are indicated on each end of the room, and the dining room's fireplace is centered on the wall adjoining the kitchen. The staircase has been repositioned from a corner to the center of the wall between the living room and dining room. The drawing of the second floor shows the space has been redesigned for the family. The room labeled "large chamber" became the girls' bedroom, and the lucky young ladies had their own bathroom. A rarity at the time, the second floor offered three full bathrooms.

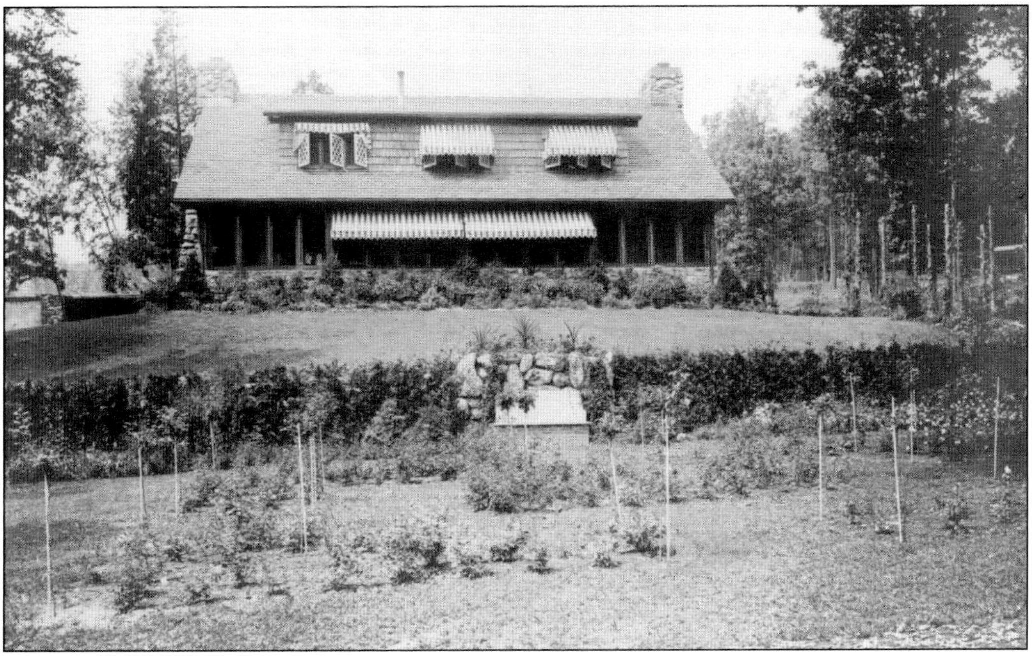

Looking up at the log house from the rose garden emphasizes the house's scale and firm connection to the surrounding land. Viewed from this angle, it does not appear to be constructed of logs. The foundation of local fieldstone spans the full width of the house. The second floor, seen from the front as a shed dormer, is frame construction faced with cypress shingles.

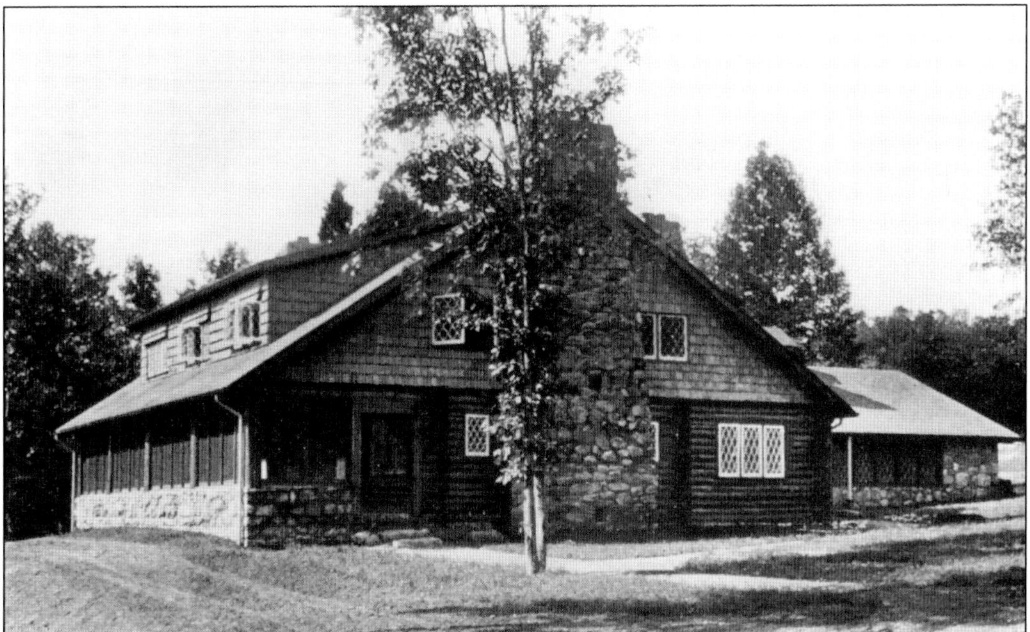

The log construction is visible from this angle. The first floor is constructed of peeled logs, which were hewn on the property and stained a dark woodsy brown to match the original tree bark. This early photograph shows the log house before the foundation plantings were installed. The original Ruberoid rolled roofing has not yet been replaced by the soft green Ludowici clay tiles.

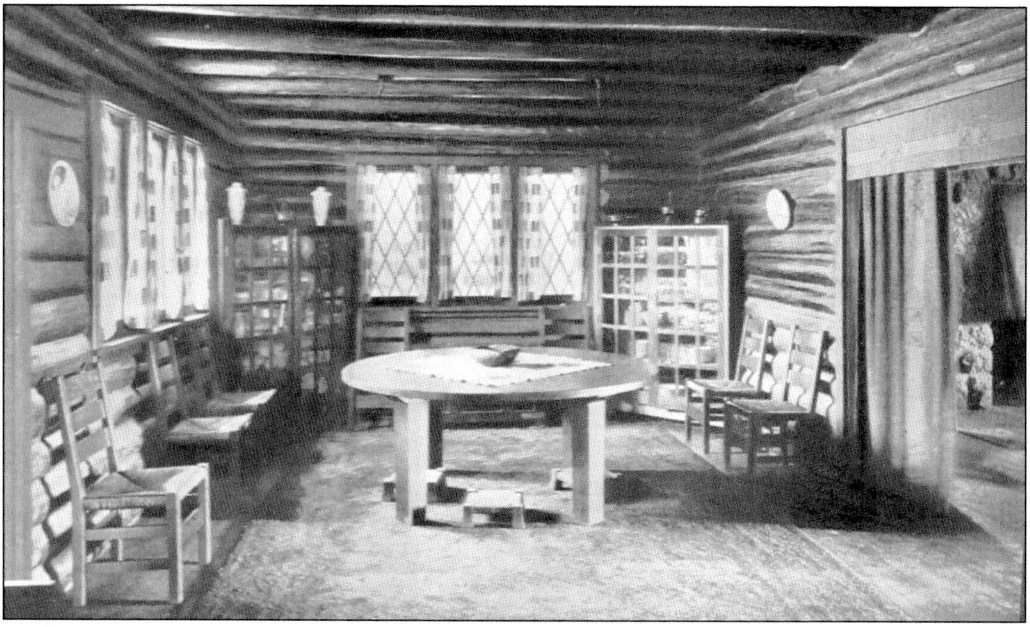

In the north end of the dining room are two oak corner cabinets fitted with copper hardware. This pair was made specifically for his dining room. Embroidered portieres soften the look of the doorway to the living room. On the floor is a woolen rug, produced by Alexander Morton and Co., featuring a sunny golden background with floral designs in pumpkin and green.

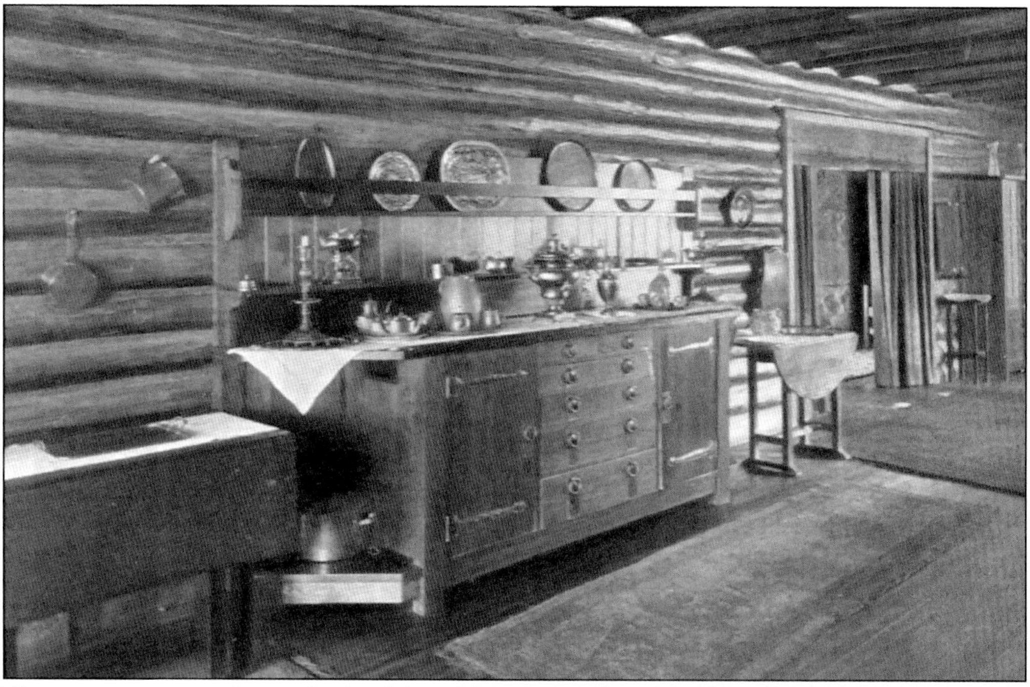

This massive custom-designed oak sideboard featuring a chamfered-board back, hammered copper pulls, and large strap hinges was the centerpiece of the "serving" area in the dining room. An embroidered linen runner sits atop the sideboard, which is filled with hammered copper, ceramic, and silver serving pieces.

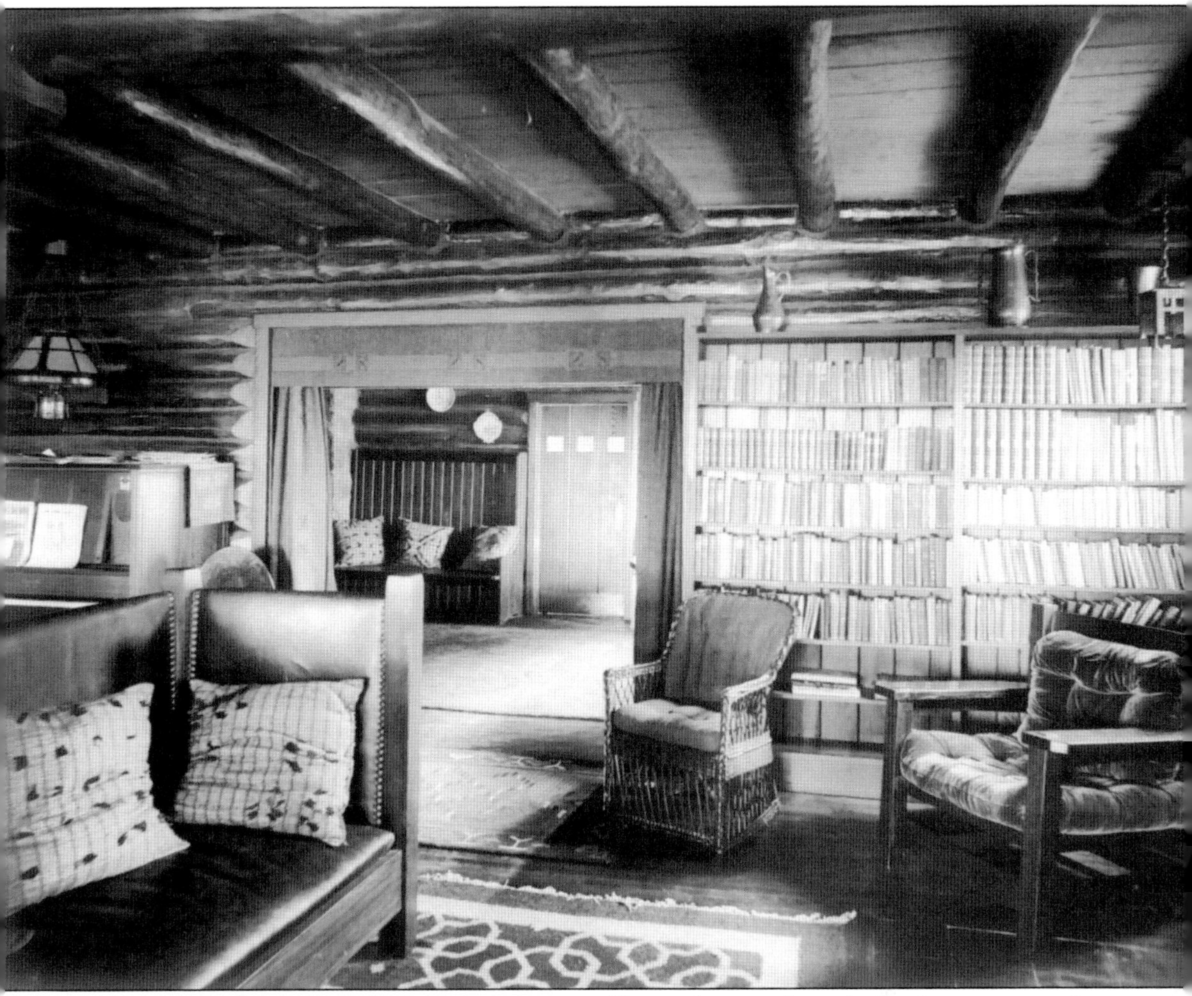

This view of the north end of the living room offers a glimpse through the doorway into the dining room. On the far left, a piano lamp, not seen in earlier pictures of the room, hangs over the upright piano. The cozy corner of the room is replete with signs of a comforting home life: the piano, the inviting leather settle, the wall of books, and the hefty Eastwood chair, which many believe was Stickley's personal favorite. *The Craftsman* editor Natalie Curtis wrote, "The house of logs appeals to us because it is a part of our heredity. It was a primitive home to man, a rudimentary sheltering of domestic life, a place of safety where love and friendship could be shut in and foe and danger shut out. . . . There is nothing about it that can remind one of street and city, it only deepens the sense of glad release from tension and artificiality."

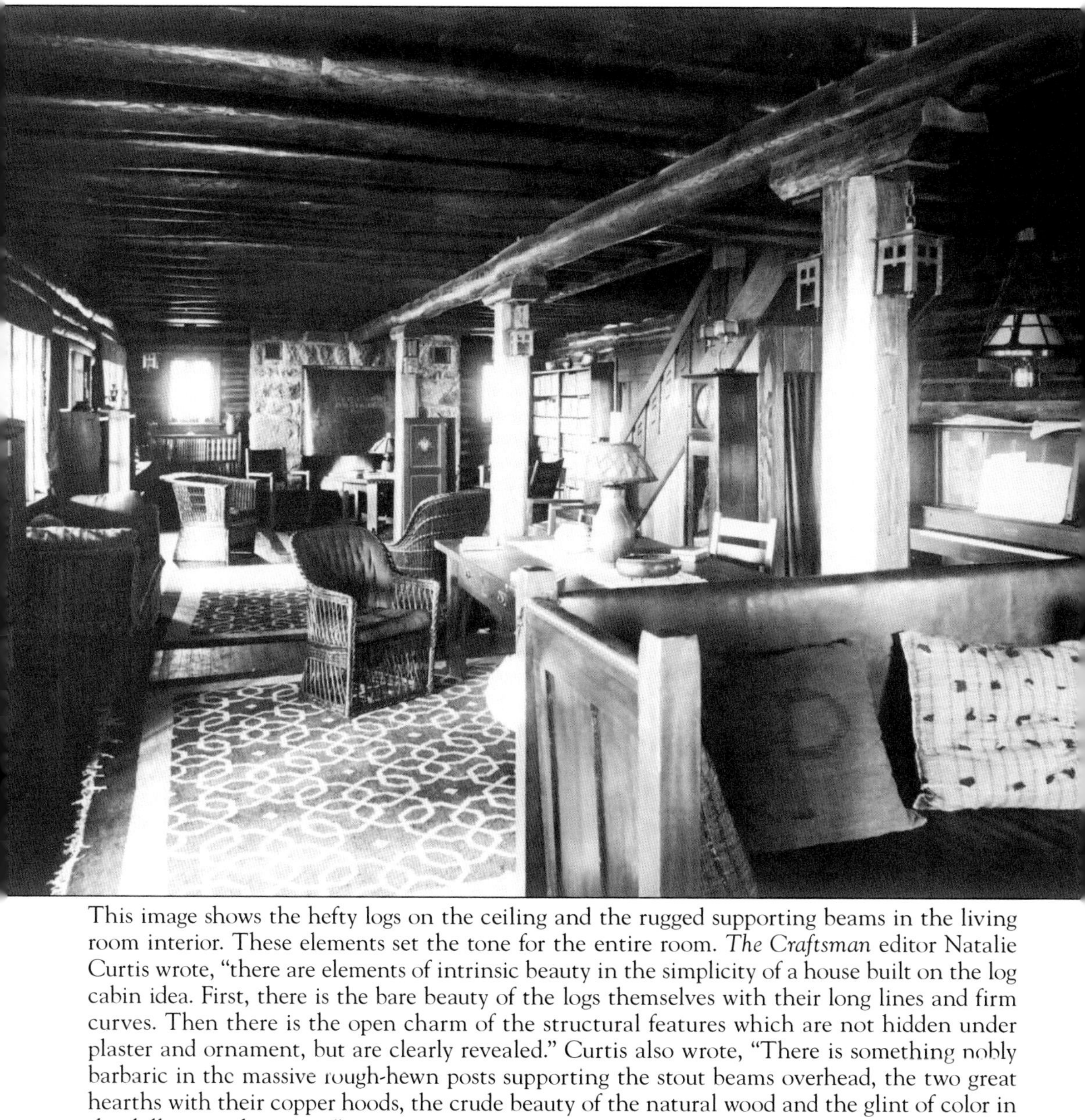

This image shows the hefty logs on the ceiling and the rugged supporting beams in the living room interior. These elements set the tone for the entire room. *The Craftsman* editor Natalie Curtis wrote, "there are elements of intrinsic beauty in the simplicity of a house built on the log cabin idea. First, there is the bare beauty of the logs themselves with their long lines and firm curves. Then there is the open charm of the structural features which are not hidden under plaster and ornament, but are clearly revealed." Curtis also wrote, "There is something nobly barbaric in the massive rough-hewn posts supporting the stout beams overhead, the two great hearths with their copper hoods, the crude beauty of the natural wood and the glint of color in the dull orange hangings."

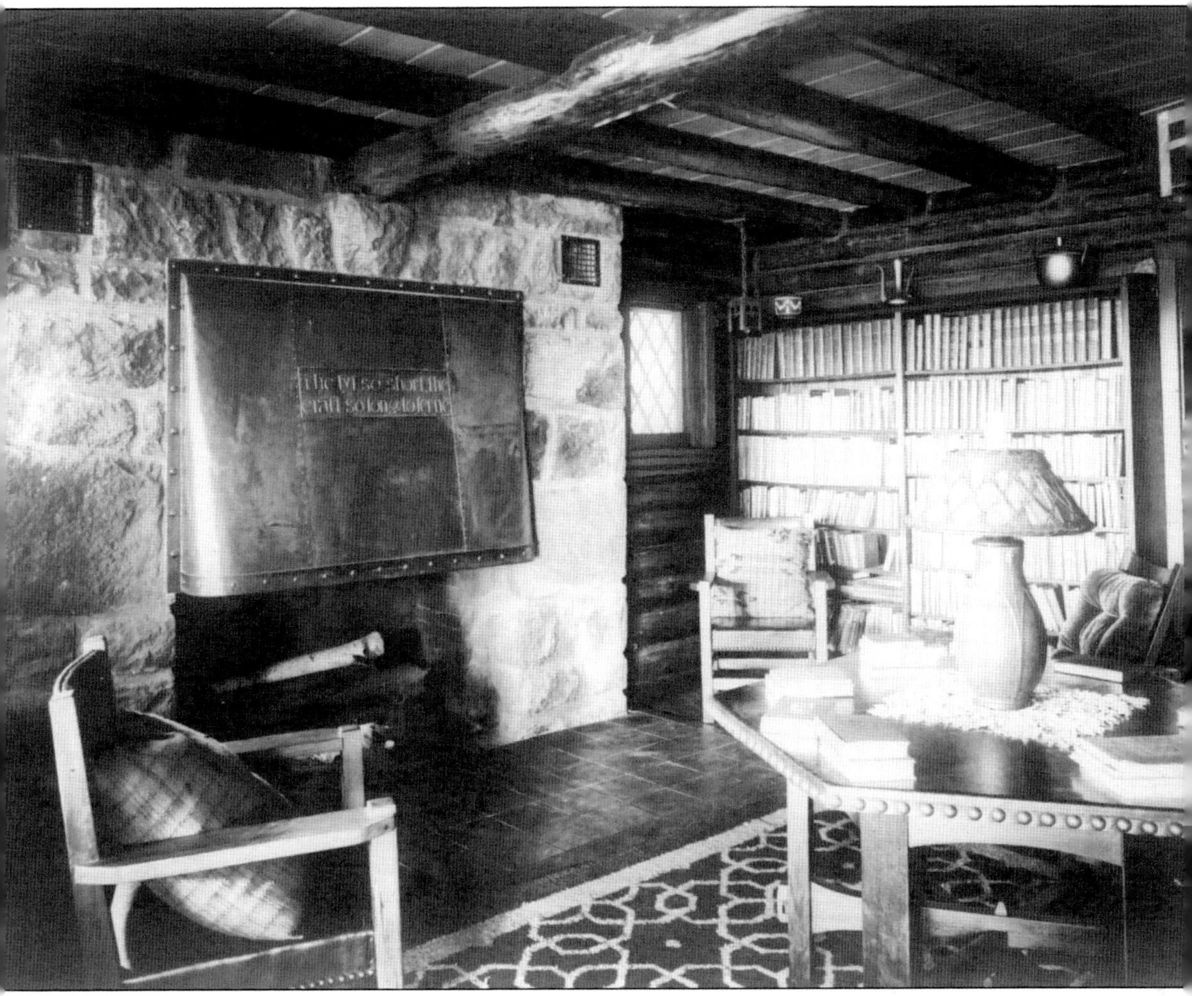

The large living room was visually divided into three zones, much like the "great rooms" of today. Seating areas were arranged near each of the room's two stone fireplaces. The photograph shows the seating area nearest the fireplace that dominates the south end of the room. All of the house's five fireplaces have mottos hammered into their metal hoods. This one quotes Geoffrey Chaucer's appropriate phrase, "The lyf so short, the craft so long to lerne." To the right, a chestnut bookcase lined with books creates a backdrop for the setting. In front of the bookcase is a hexagonal table with leather top anchored in place by round-headed brass tacks, which echo the pleasing rhythmic patterns on the chair tacks as well as the fireplace rivets. Natalie Curtis wrote, "The new log house at Craftsman Farms . . . is a log cabin idealized. Though adapted to the demands of modern civilization it yet has the charm of the more primitive dwelling and creates the same suggestion of simple and natural living and kinship with the outdoor world."

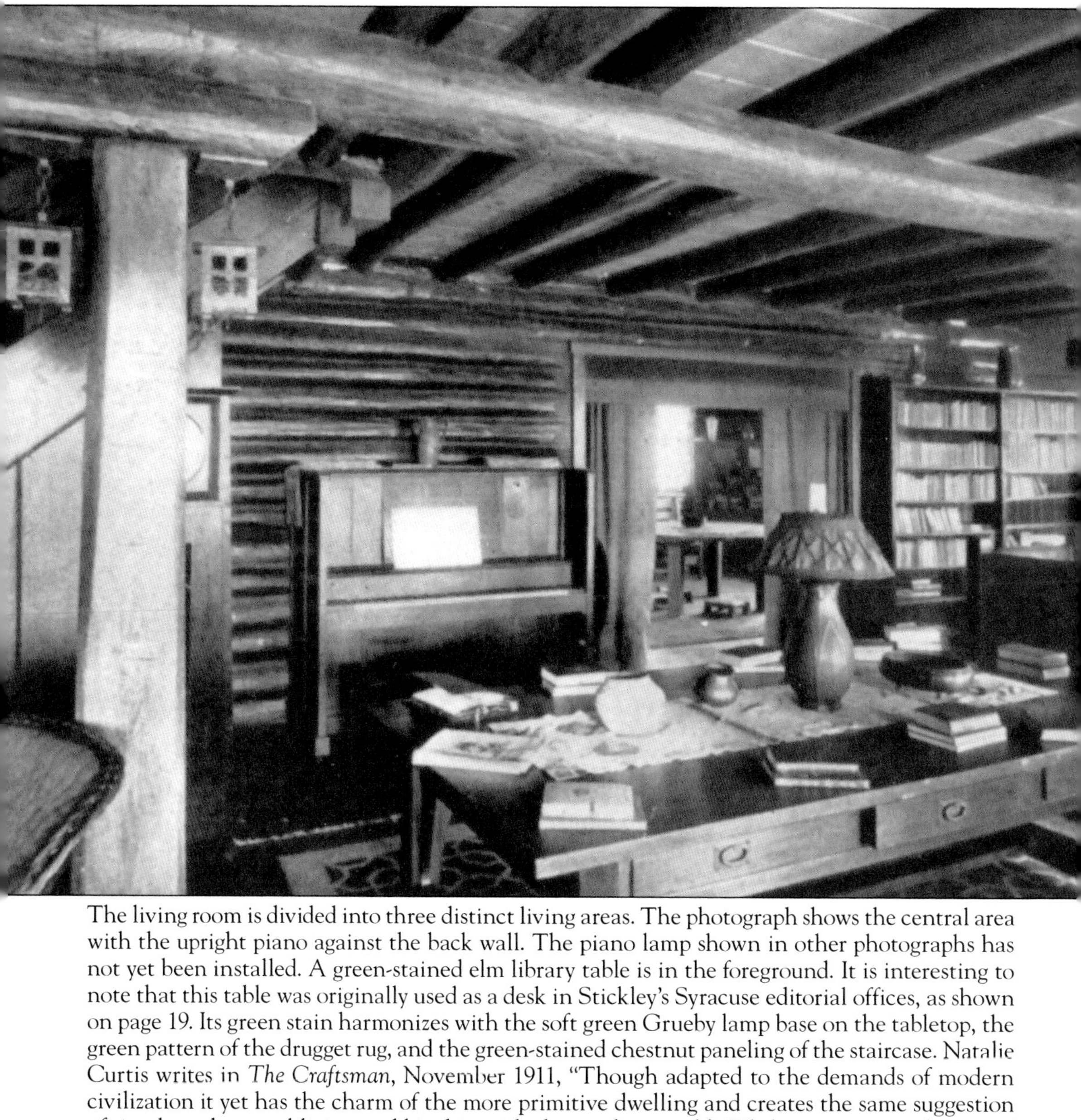

The living room is divided into three distinct living areas. The photograph shows the central area with the upright piano against the back wall. The piano lamp shown in other photographs has not yet been installed. A green-stained elm library table is in the foreground. It is interesting to note that this table was originally used as a desk in Stickley's Syracuse editorial offices, as shown on page 19. Its green stain harmonizes with the soft green Grueby lamp base on the tabletop, the green pattern of the drugget rug, and the green-stained chestnut paneling of the staircase. Natalie Curtis writes in *The Craftsman*, November 1911, "Though adapted to the demands of modern civilization it yet has the charm of the more primitive dwelling and creates the same suggestion of simple and natural living and kinship with the outdoor world. While retaining its dignity of architecture it invites a sense of informality, of intimacy. Here the visitor feels instinctively and immediately at home."

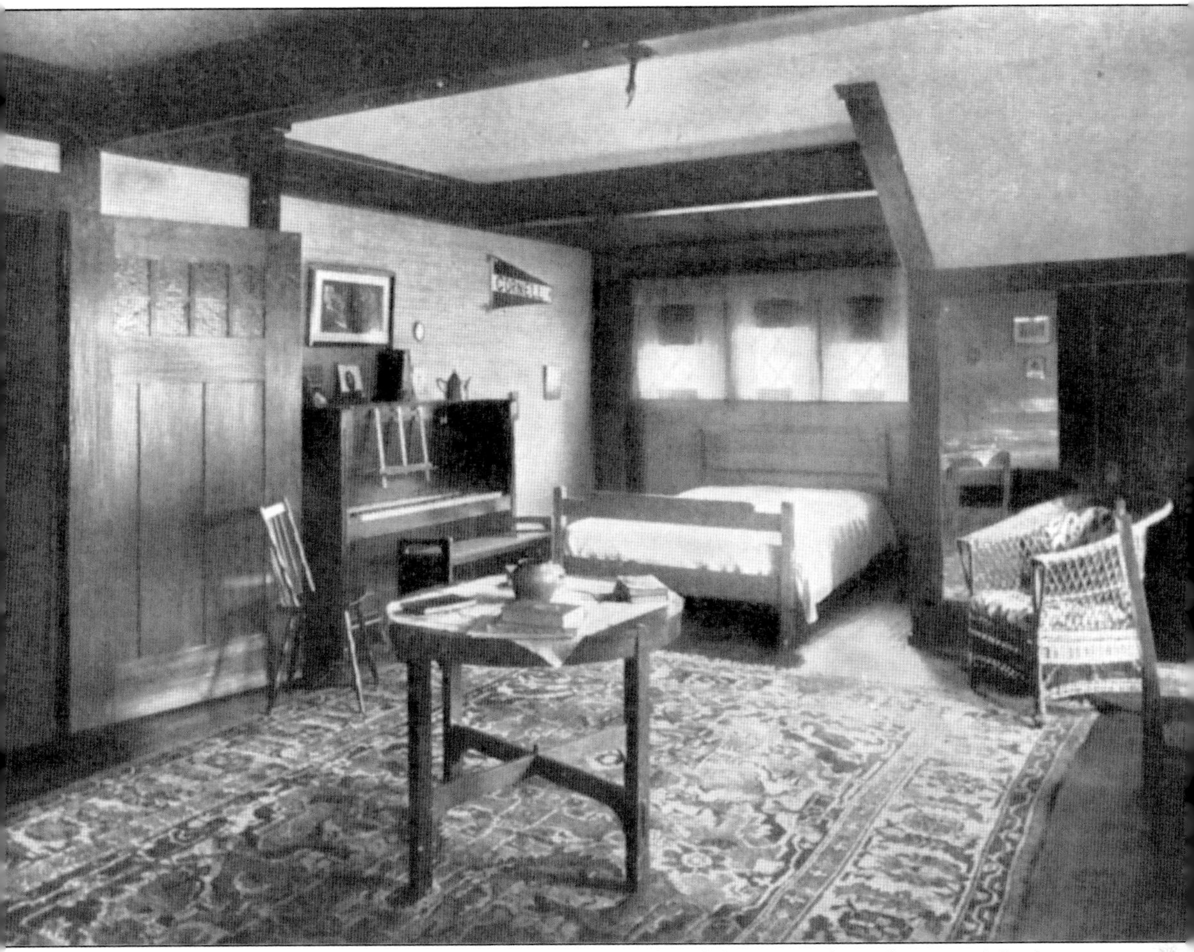

The girls' bedroom, shown here, is much more refined and feminine than the rest of the home. Natalie Curtis described it best in *The Craftsman*, November 1911: "This latter room has a peculiar charm because of its woodwork of dark gumwood, which is perhaps as beautiful as any produced by Craftsman design. The color scheme has a certain feminine refinement. The walls are covered with gray Japanese grass-cloth and the hearth is of dull blue Grueby tiles with a brass hood. The furniture is gray oak, decorated with a small design inlaid with blue and copper. This design is outlined in black and is crossed by a delicate vine-like figure of greenish yellow. Unlike those bedrooms in which 'daintiness' is expressed by weakness, tawdry trimmings, flippancy and ruffles, this room has both delicacy and strength and is thus appropriate to the ideal of the modern woman."

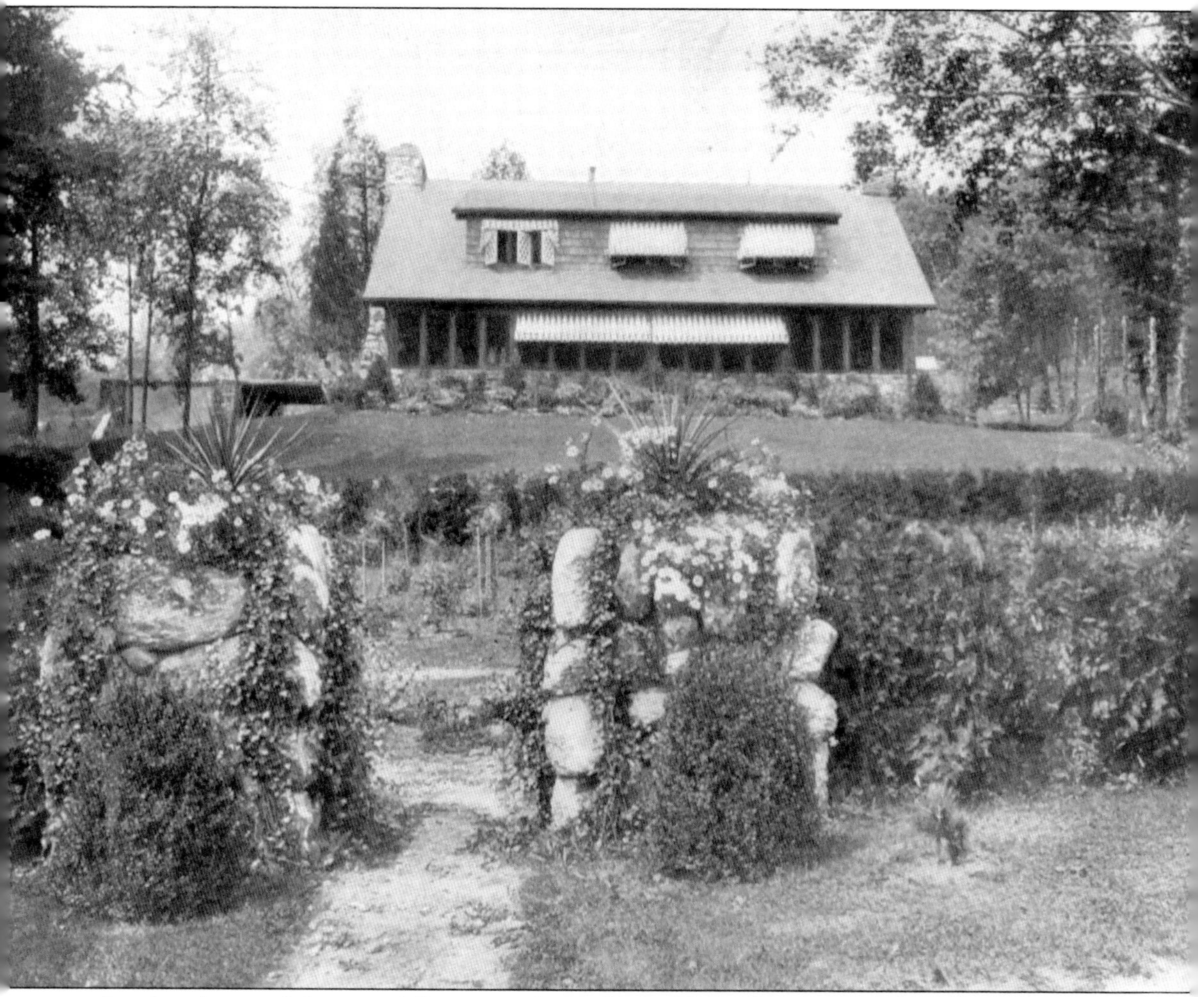

An abundance of plants and flowers exuberantly overflows the stone piers in this photograph. Between the piers a pathway leads to the sunken rose garden. A lawn rises above the stone wall that edges the rose garden, leading the eye up to the house. The sturdy stone foundation supports the log house porch, linking harmoniously to the stone walls, stone piers, and the very ground from which these stones were dug. Above the porch windows, an awning offers shade and an invitation to linger on the porch that Natalie Curtis described as "swept by every passing breeze, nestling amid the green of the hillside . . . like a vast outdoor living room." She concluded her 1911 essay on Craftsman Farms by saying, "It charms the visitor by its harmony with nature and its unity of the best in civilization with the best in cruder forms of life. To the strength, the courage and the honest effort typified by the primitive log cabin, art has here added the grace of beauty, and science the requisites of comfort."

# *Four*

# The Stickley Years

Three cottages at Craftsman Farms were completed in the spring of 1910. The family left Syracuse to live in one of them while the main house was constructed. Eldest daughter Barbara, who was 23 at the time, had already left their Syracuse home. After attending Smith College, she went to work at her father's offices, and when the family moved to Craftsman Farms, she was sharing her father's New York City apartment on West Ninety-Seventh Street. The other children, Mildred, 22; Hazel, 19; Marion, 17; Gustav Jr., 15; and Ruth, 13, settled into life at Craftsman Farms.

Little is known about the time the handsome young Gustav Jr. spent at Craftsman Farms. Stickley scholar David Cathers notes that he "resisted his father's efforts to initiate him into the rugged, rural life advocated by *The Craftsman* magazine." Instead, Gustav Jr. attended a military academy and then joined the US Navy.

It is hard to imagine what the girls and their mother felt about leaving the social life of Syracuse for a farm in rural New Jersey. Little is known about the thoughts of Eda. She had been a seamstress prior to her marriage to Gustav and enjoyed spending her free time knitting and doing needlework. She was also a loving mother who had close relationships with her children.

It was not long before the lively daughters made plenty of friends in their new community. The family moved into the main house in 1911, and almost immediately, the house was the site of social events. The July 28, 1911, edition of the local newspaper, the *Jerseyman*, featured an article entitled "The Misses Stickley Entertain," describing a dance party at Craftsman Farms attended by "sixty young people of Morristown and Morris Plains."

Marion Stickley's scrapbook contains many dance cards, photographs of fun times, and a poem written by an admirer named John W. Bloodgood. Entitled "The New Morris Plains," it compares Morris Plains before the Stickley girls arrived to the community after their arrival. In charming rhyme, it claims that "a stagnant lot we used to be" until the arrival of the Stickley girls, who brought about "good times."

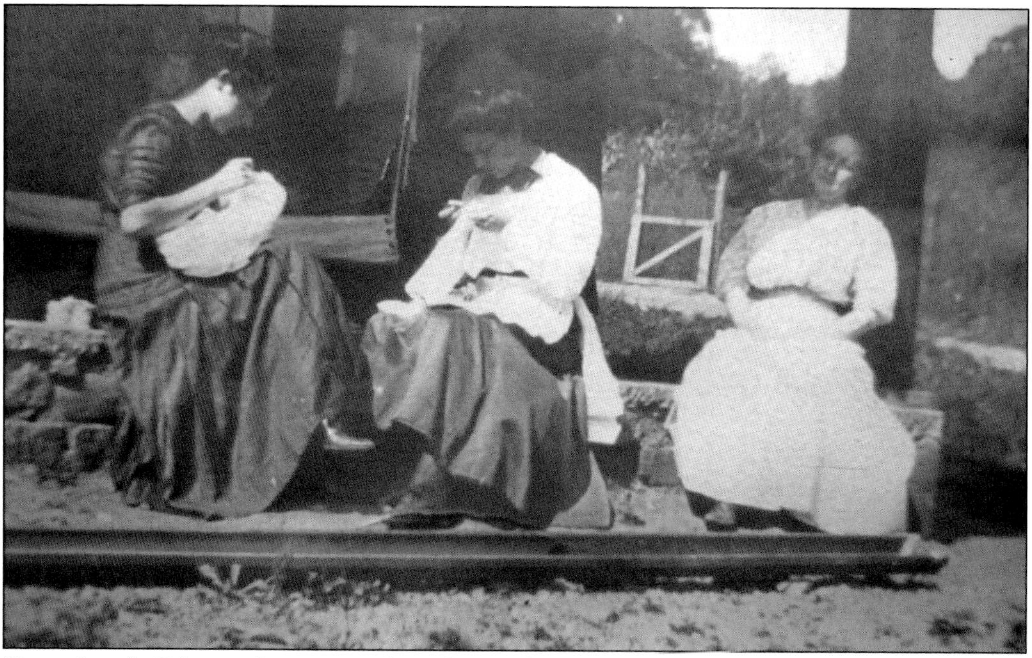

Gustav's wife, Eda, can be seen on the right on the open porch of one of the cottages. The two girls are presumed to be daughters. The cottages, which look as if they were still partially under construction in this photograph, were the first buildings to be completed at Craftsman Farms. The family lived in a cottage briefly while the main house was constructed. (Courtesy the Stickley Museum at Craftsman Farms, Robert and Linda Preim Collection.)

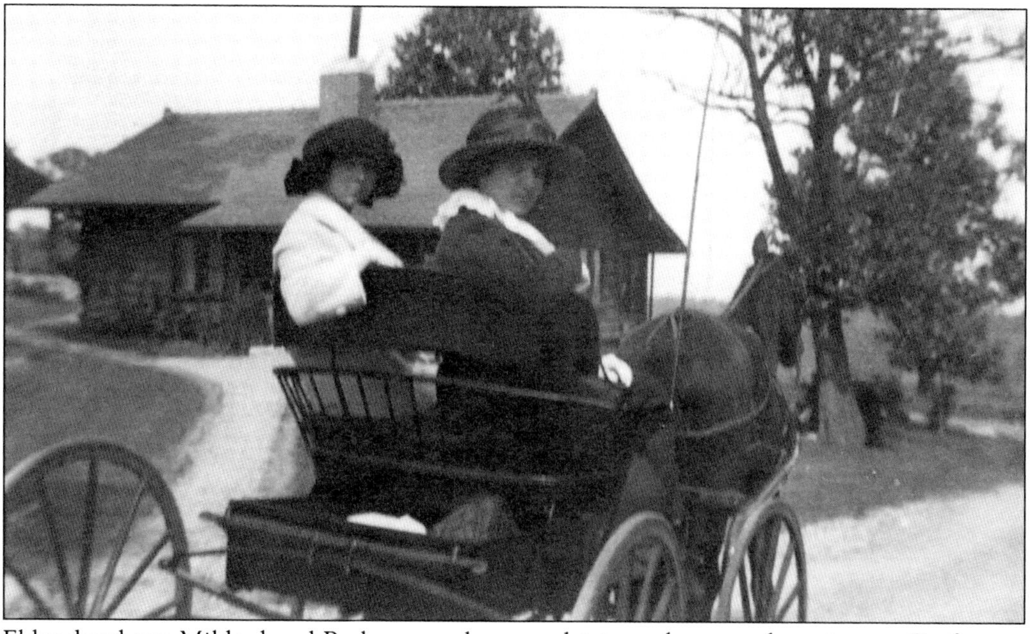

Elder daughters Mildred and Barbara can be seen driving a horse and carriage at Craftsman Farms. The south cottage is to the left and appears to have been completed. The girls, dressed in coats, hats, and gloves, have parcels stowed in the back of the carriage and may be returning from a shopping trip in Morristown. (Courtesy Cynthia McGinn.)

All the Stickley girls had reputations for being fun-loving, lively, and popular. Two younger daughters, Ruth, on the left, and Marion, who is dressed as a boy, can be seen here dancing outdoors at Craftsman Farms. Perhaps they were practicing their dance steps for one of the many parties that took place at Craftsman Farms. (Courtesy the Stickley Museum at Craftsman Farms, Robert and Linda Preim Collection.)

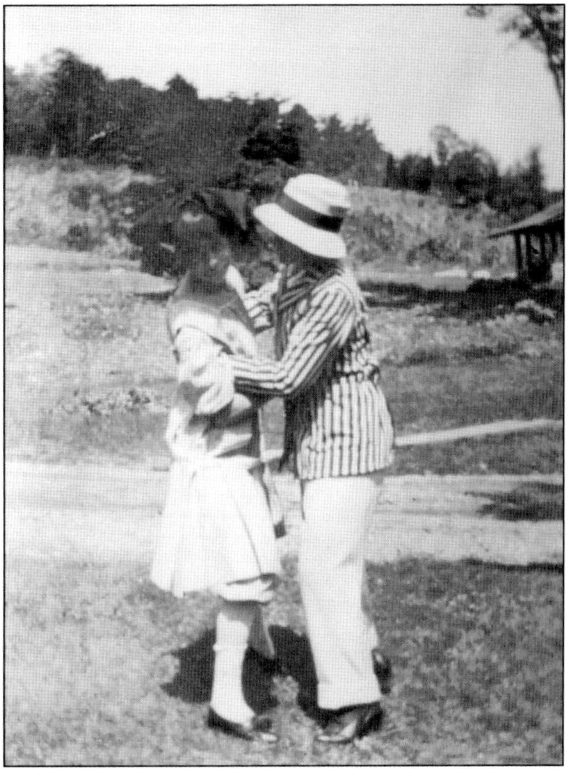

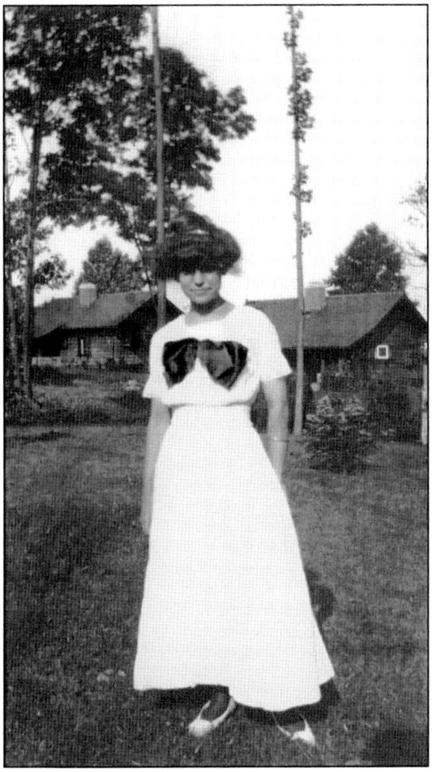

Hazel Stickley was 19 years old when the family moved to Craftsman Farms. She is seen here a year or two later, wearing the Gibson girl look that was so fashionable at the time. The cottages can be seen behind her and are all finished, complete with landscaping and modern outdoor lighting. (Courtesy the Stickley Museum at Craftsman Farms, Robert and Linda Preim Collection.)

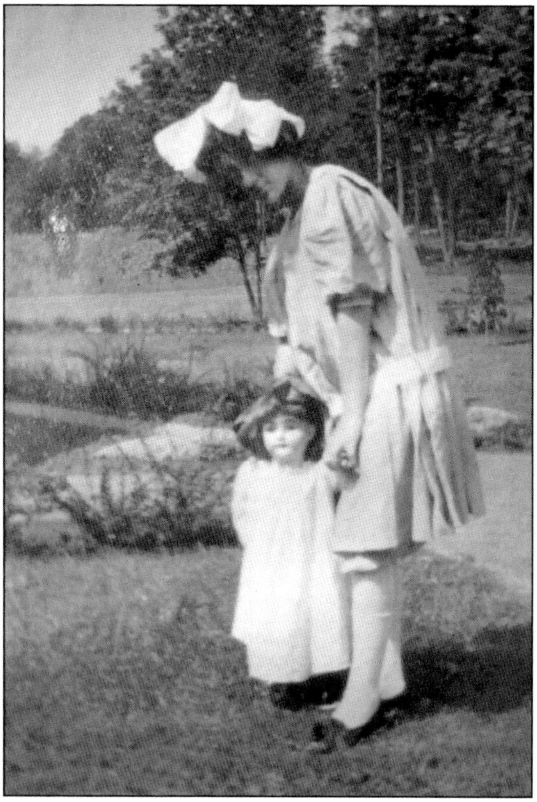

Stickley's daughter Marion was 17 when the family moved to Craftsman Farms—well beyond the age when girls play with baby dolls. She was the liveliest of the daughters and known for her sense of humor. It is likely that she dressed up like a little girl with a big bow in her hair and posed with a doll taking a walk just for the fun of it. She wears the same outfit standing beside her mother, Eda, who is dressed in a simple and conservative long linen dress. Eda was known to be a quiet and somewhat shy woman who probably felt a bit uncomfortable with her husband's fame. (Both, courtesy the Stickley Museum at Craftsman Farms, Robert and Linda Preim Collection.)

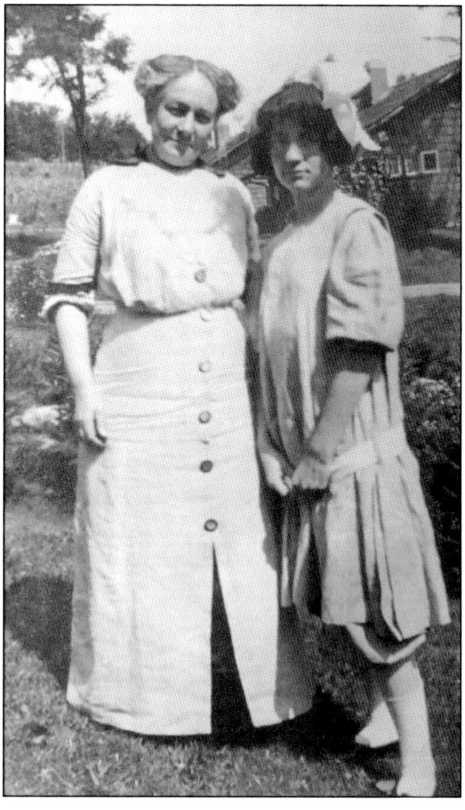

Eldest daughter Barbara met Ben Wiles at her father's firm, where Wiles worked as circulation manager for *The Craftsman* magazine. Barbara and Ben were married in October 1911 on the log house porch at Craftsman Farms. The wedding, which had over 100 guests, was written up in the *Jerseyman*. Barbara can be seen here posing in her wedding dress in one of her father's willow chairs. The couple lived briefly in New York but soon settled into one of the cottages. Below, Barbara is shown in one of the cottages wearing a nightcap and holding her daughter, also named Barbara, who was born in August 1912. (Both, courtesy Barbara Fuldner.)

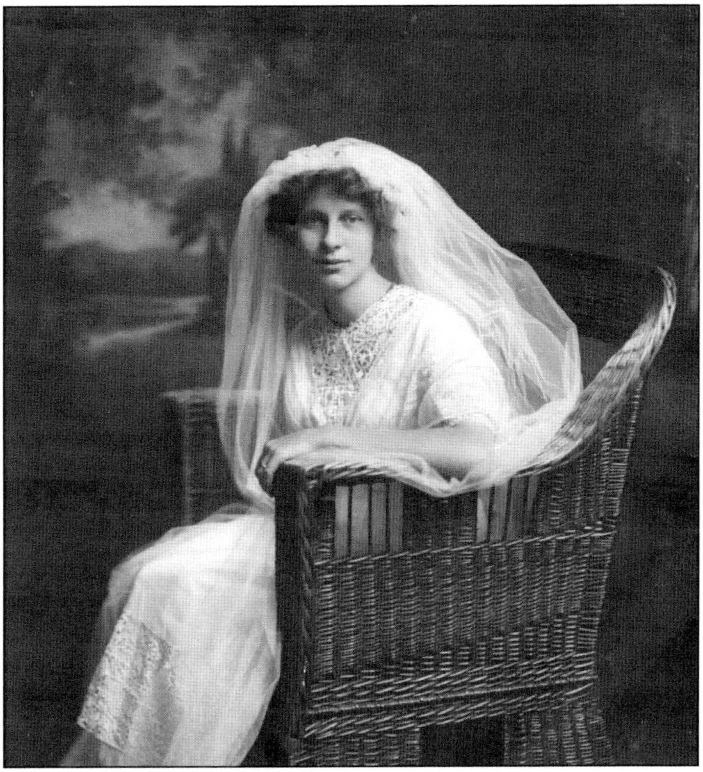

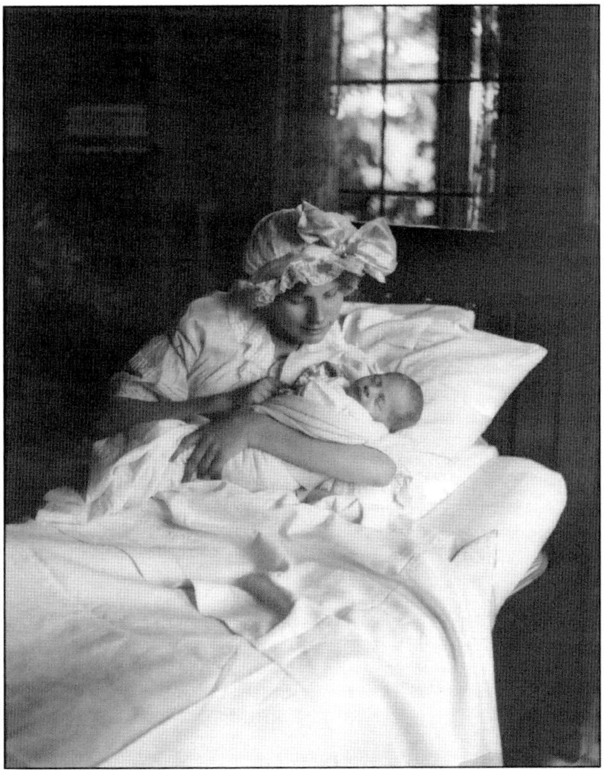

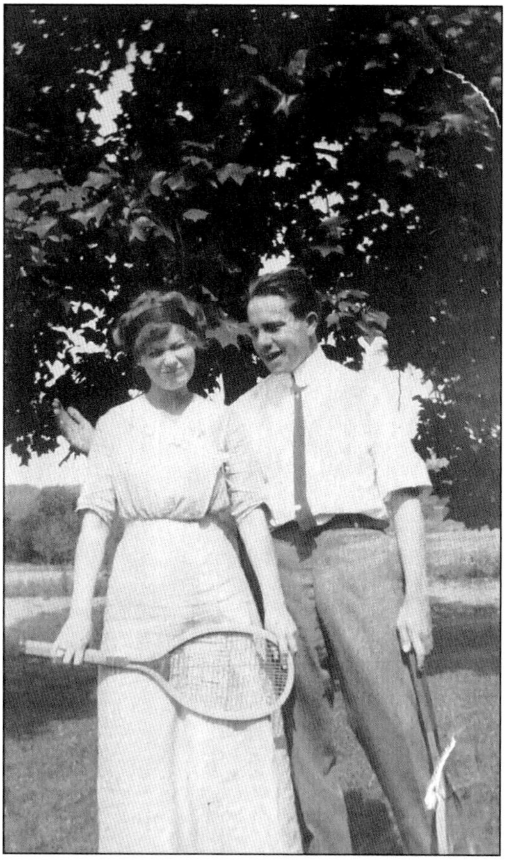

In this pre- or post-tennis photograph, Gustav Stickley Jr. has his arm around an unidentified girl. He enjoyed some time at Craftsman Farms despite a bit of continuing conflict with his father. Stickley was strict with Gustav Jr., who did not share his father's love for Craftsman Farms or the rugged character-building work his father wanted for him. Having spent his formative years as the son of a well-to-do businessman, his world was quite different than that of his father. Young Gustav was closer to his mother, and they apparently had a loving relationship. Below is a photograph believed to be of his sister Marion, dressed in lace-up shoes and a white linen dress, as if she, too, had been playing tennis. (Left, courtesy the Stickley Museum at Craftsman Farms, Robert and Linda Preim Collection; below, courtesy Cynthia McGinn.)

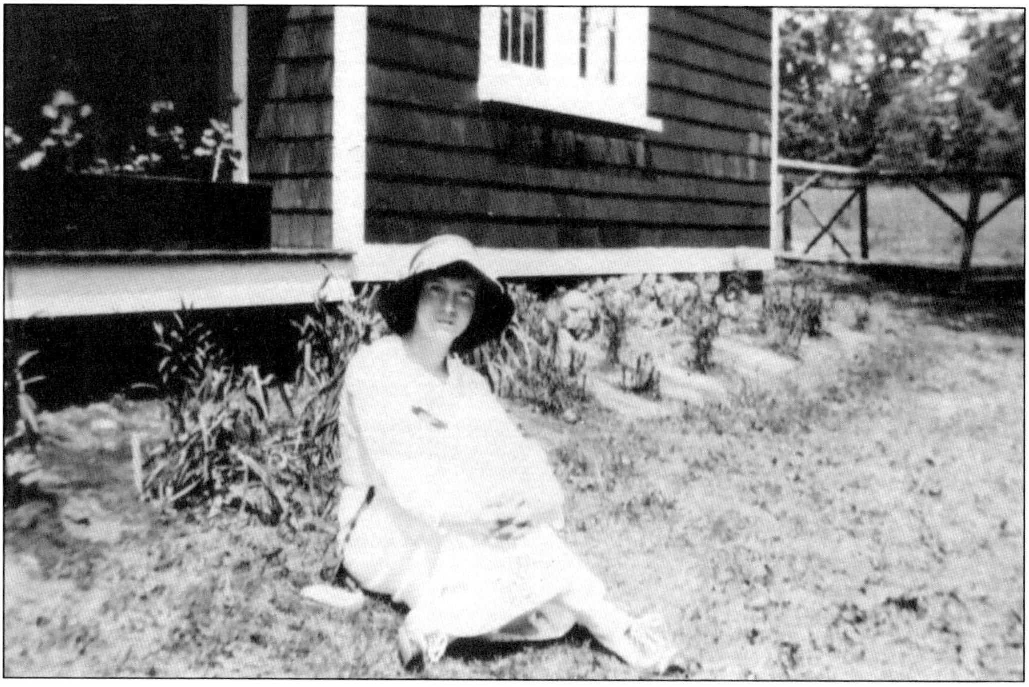

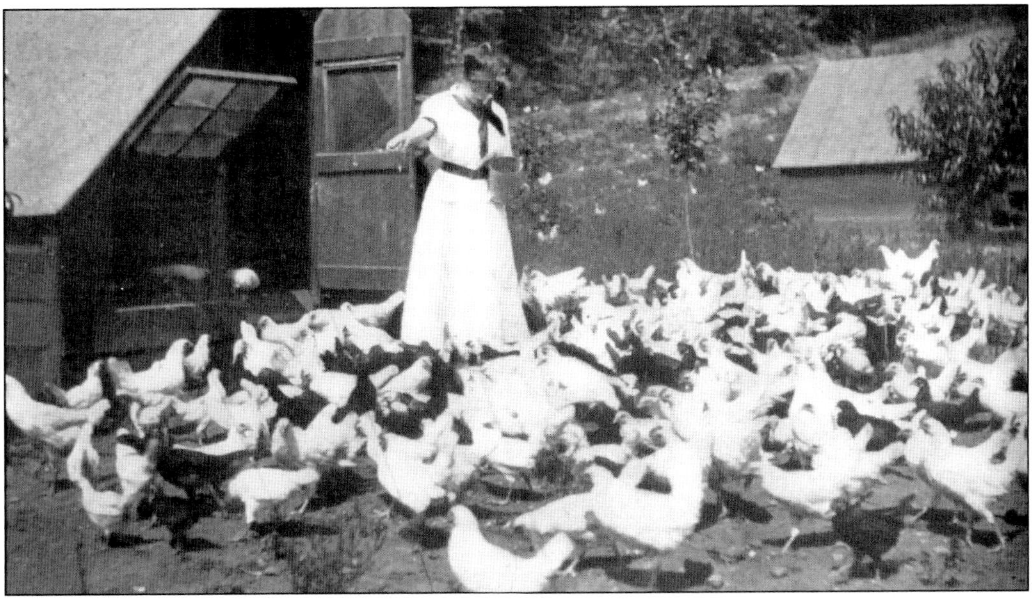

This photograph of daughter Mildred feeding the chickens at Craftsman Farms was published in *The Craftsman*, November 1912. Mildred was not really responsible for the chickens but was a hard worker with many chores. She married George Cruess in September 1915 in the garden at Craftsman Farms. She and her husband then moved into one of the cottages.

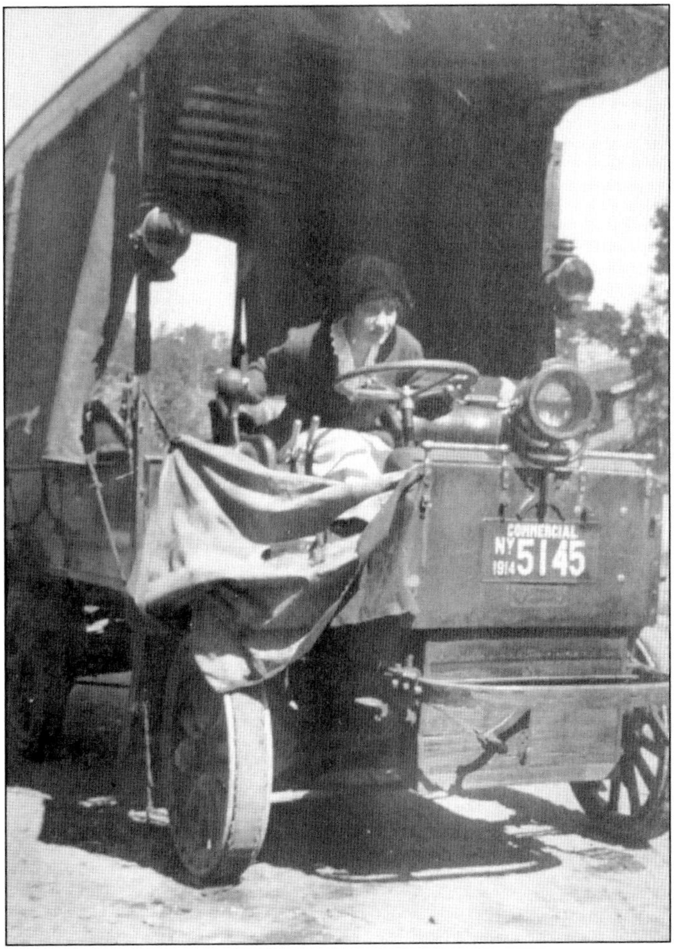


Daughter Marion is playfully acting out honking the horn and steering a large truck. The New York license plate indicates it may have been one of the electric trucks that delivered produce to the Craftsman Restaurant. Marion married Carl Preim in June 1916 on the log house porch. She and Carl moved into a cottage on the property. (Courtesy the Stickley Museum at Craftsman Farms, Robert and Linda Preim Collection.)

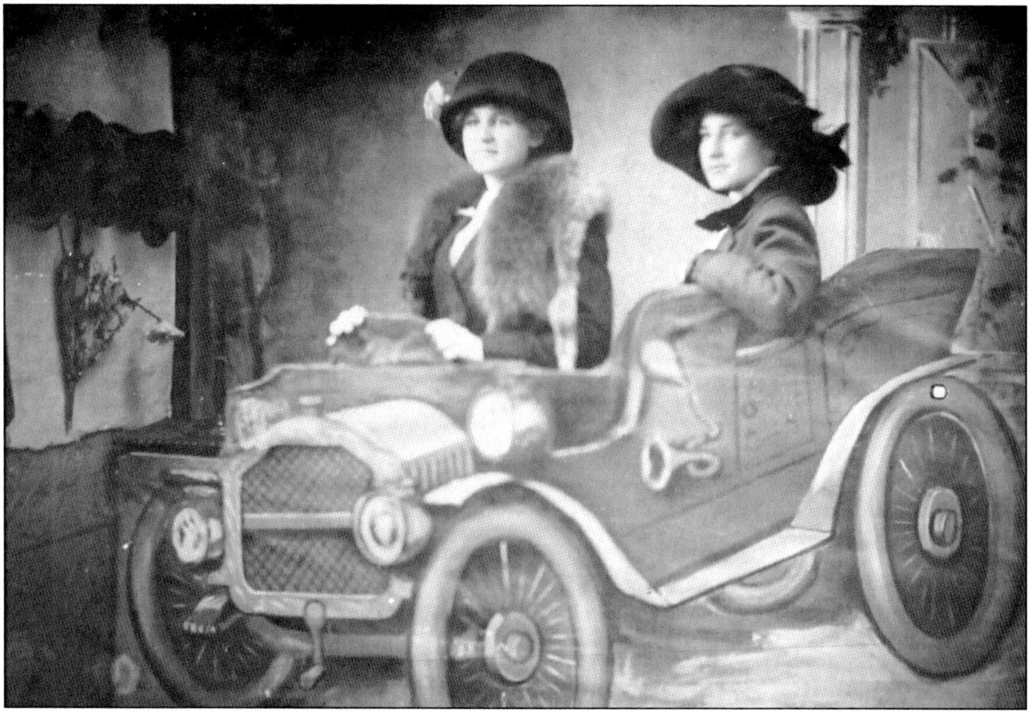

Despite living in a rustic setting, the family's affluence was apparent. The daughters made frequent trips to Manhattan for activities like lunch at the Waldorf. Here, Mildred and her sister Marion are seen in a fun cutout photograph popular at that time. Mildred is wearing what she called the "beautiful furs" her father had bought for her in November 1911. (Courtesy the Stickley Museum at Craftsman Farms, Robert and Linda Preim Collection.)

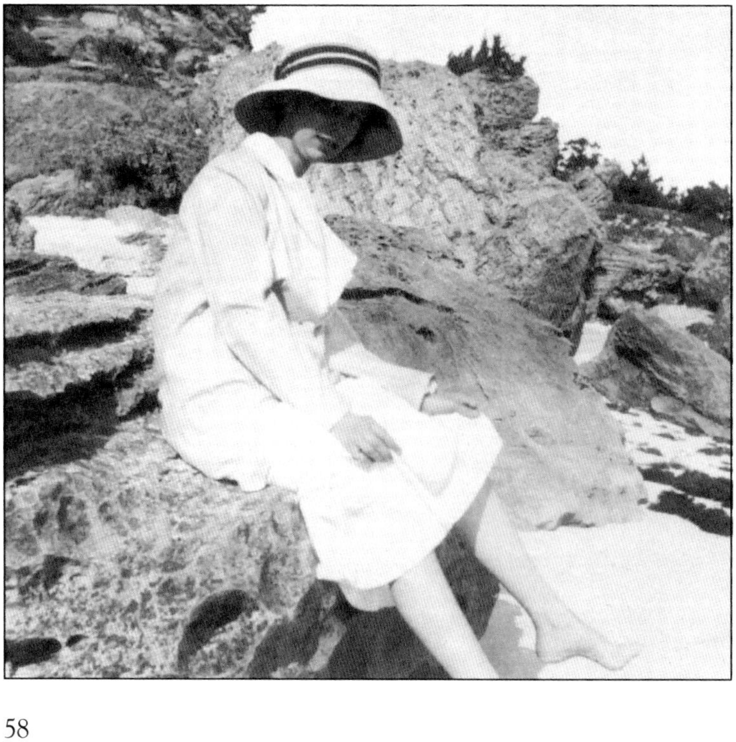

A fine example of the family's prosperity could be seen in January 1913, when Eda and three of her daughters went on a cruise to Bermuda. Hazel is shown here barefoot on the Bermuda beach with a broad-brimmed hat to shade her from the bright sunshine. (Courtesy Cynthia McGinn.)

Stickley's younger daughters are playfully pulling up their skirts to enjoy the waves on the Bermuda shore. It must have been glorious to leave a cold January in New Jersey for the warm beach and fun in the sun. The success of their father's business ventures made it possible. (Courtesy the Stickley Museum at Craftsman Farms, Robert and Linda Preim Collection.)

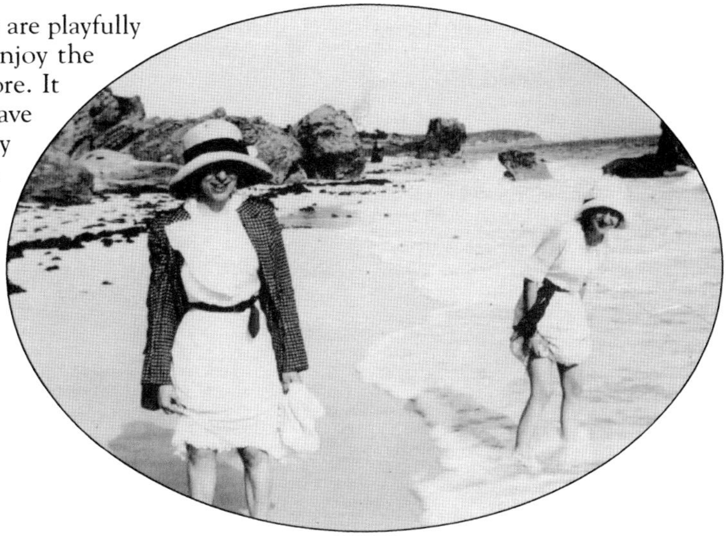

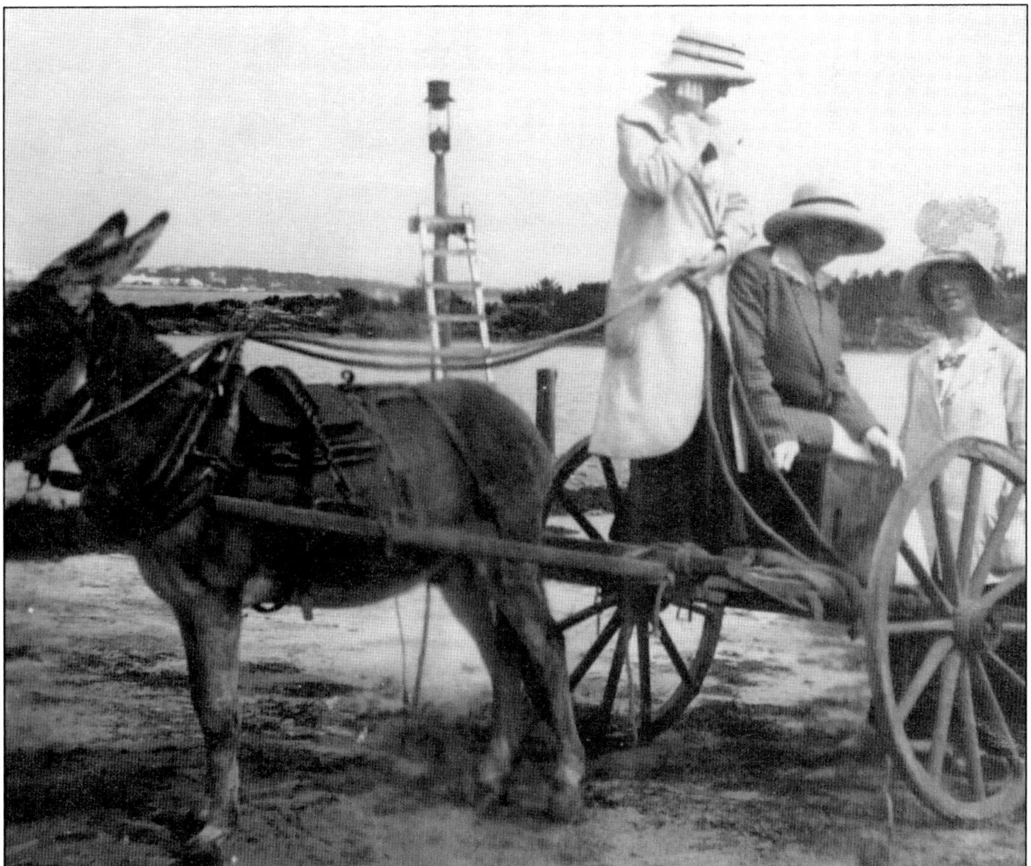

Daughters Mildred, Marion, and Hazel are stylishly dressed in tailored clothes and brimmed hats as they cavort in a donkey cart on vacation in Bermuda in 1913. Such rides were poplar attractions at beaches at that time, and the lively Stickley girls surely would not miss out on an opportunity for a fun time. (Courtesy Cynthia McGinn.)

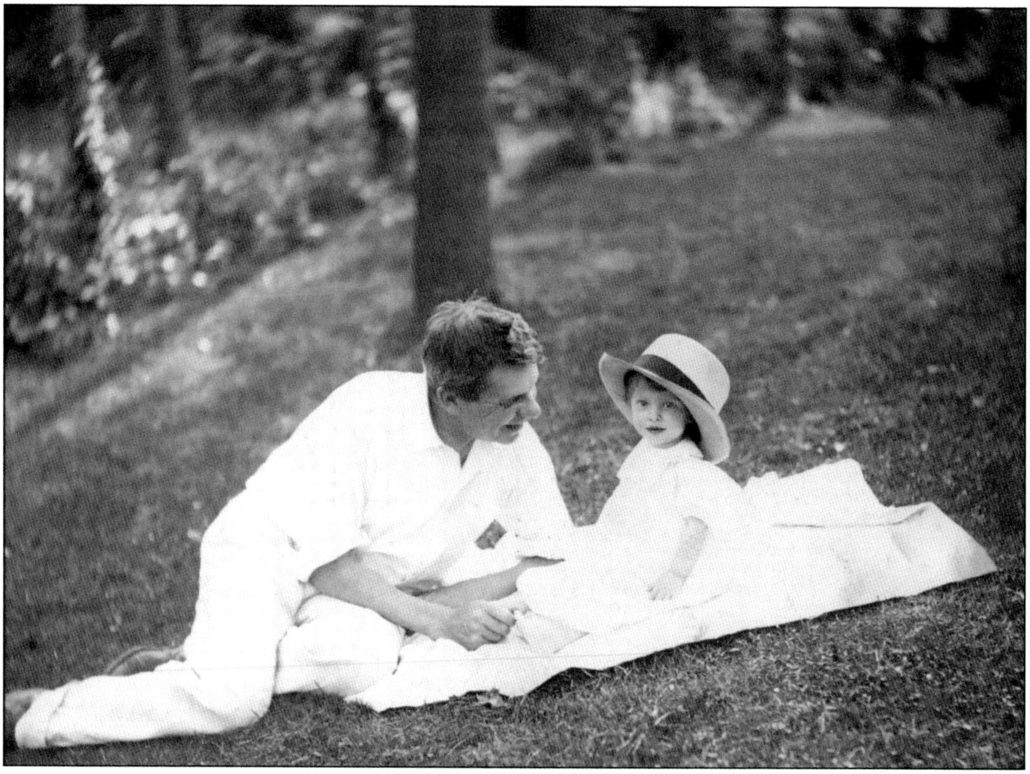

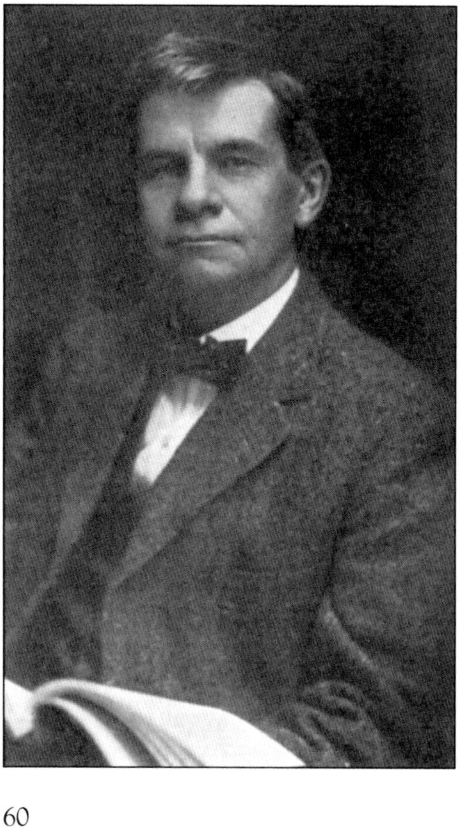

In this tender photograph, a rare glimpse of Gustav Stickley as a family man enjoying his private life is seen. He is spending time with his first granddaughter, Barbara Wiles, on the grass at Craftsman Farms. She has playfully donned his hat and smiles winningly for the camera. (Courtesy Barbara Fuldner.)

The tide was beginning to turn for the financially overextended Stickley, shown here in a portrait from the Craftsman Restaurant menu. The enormous rent of the Craftsman Building, the economic impact of World War I, and the changing tastes of the American public all combined to spell doom for his business empire. On March 23, 1915, he filed for bankruptcy. (Courtesy Barbara Fuldner.)

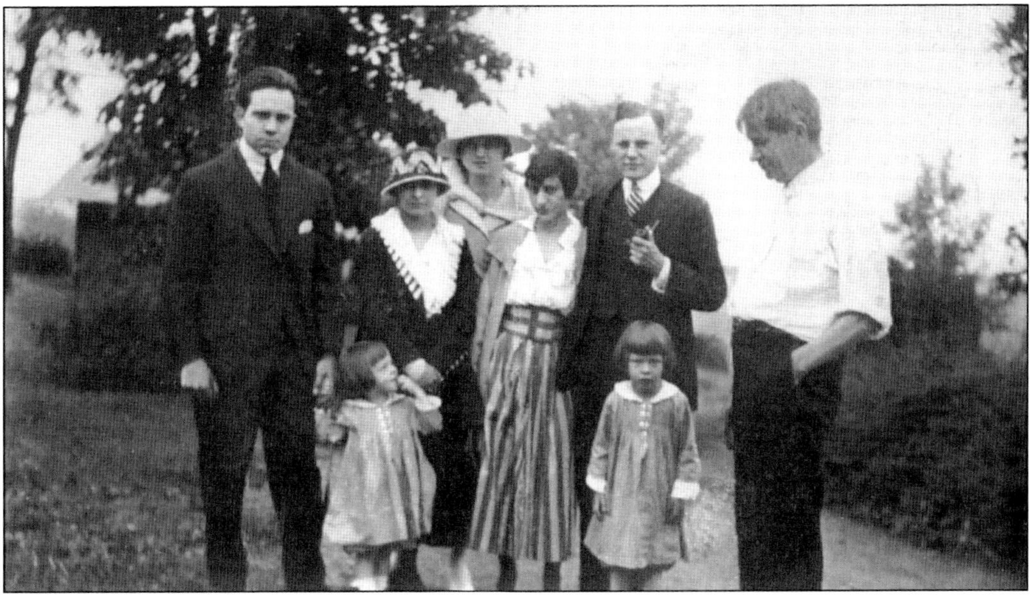

Stickley published his magazine through December 1916 and struggled to breathe life back into his businesses. This photograph was among the last taken before Craftsman Farms was sold in August 1917. Adults from left to right are Gustav Jr.; Mildred; unidentified; Marion; Marion's husband, Carl; and Gustav Stickley. Granddaughters Edith (left) and Barbara are in front. (Courtesy the Stickley Museum at Craftsman Farms, Robert and Linda Preim Collection.)

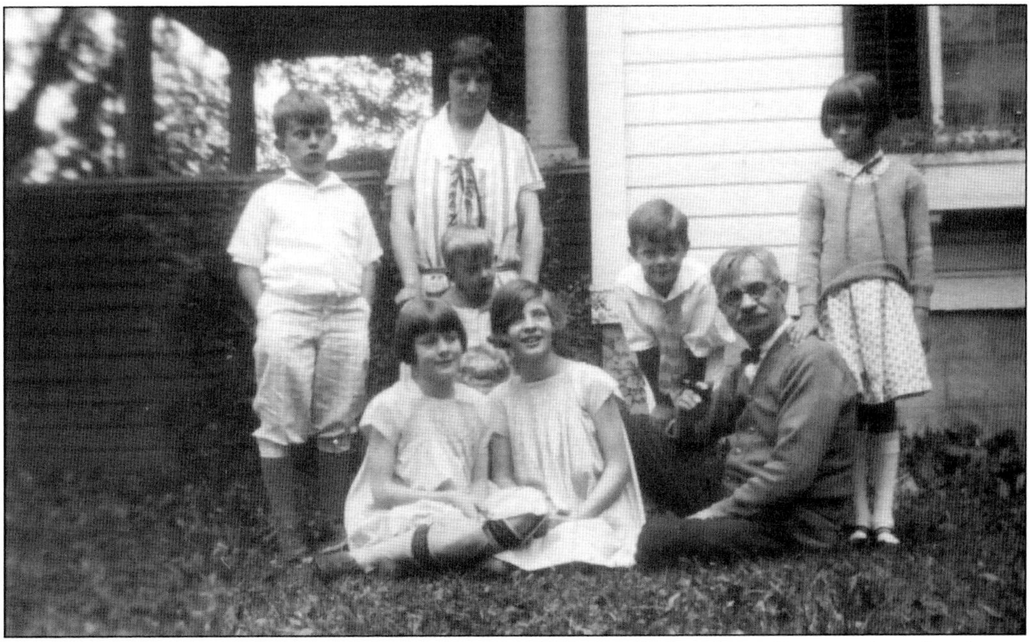

Leaving his beloved Craftsman Farms, Gustav and Eda briefly lived in a Syracuse boardinghouse called the Wolcott. That year, Eda suffered a paralyzing stroke. Daughter Mildred took care of Eda until her death in 1919. Mildred is standing in this early 1920s photograph, taken in Skaneateles Lake, New York. Stickley is seated with, from left to right, grandchildren Edith, Dick, Barbara, and Ben. Grandchildren John and Ruth are standing. (Courtesy Cynthia McGinn.)

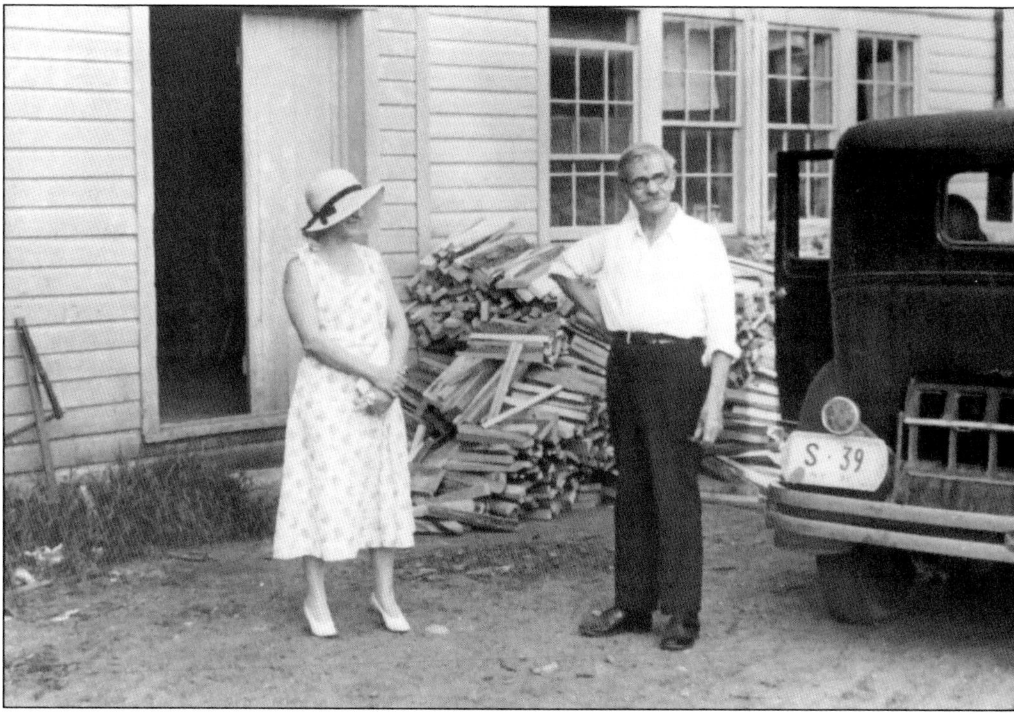


This 1930s photograph is believed to be of Stickley with daughter Marion, the only one of his children who did not move back to New York State. When Stickley left the Wolcott boardinghouse in 1918, he rented the upstairs of a house at 832 Summer Avenue in Syracuse. Hazel, Ruth, and Gustav Jr. moved in with him for a while.

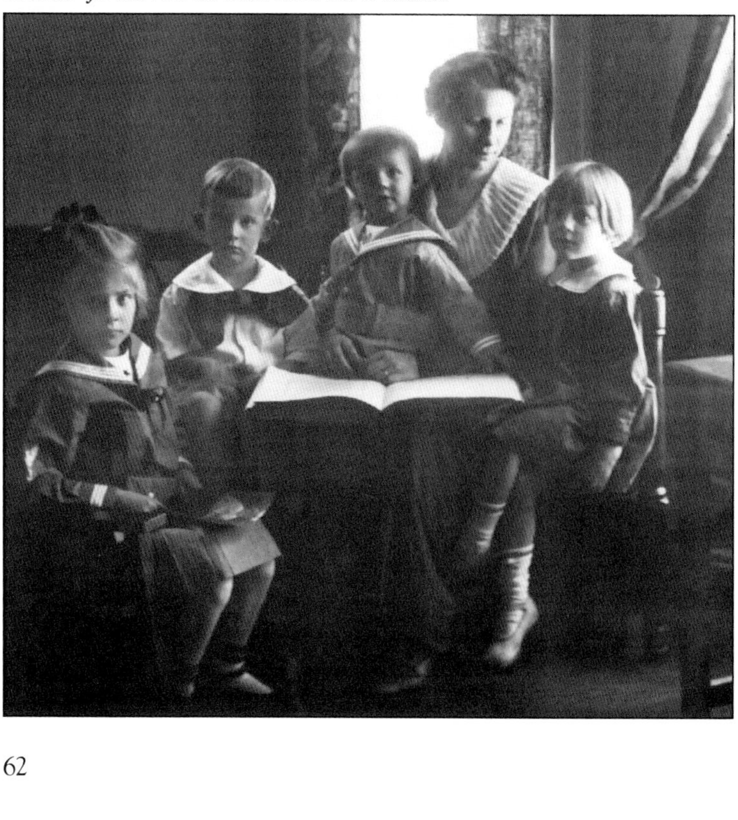

Stickley's daughter Barbara Wiles is shown here with four of her children, from left to right, Edith, John, Ben, and Barbara. Barbara and her husband, Ben, bought the old family house on Columbus Avenue in Syracuse in 1919. Stickley lived with this family on Columbus Avenue for 23 years until his death on April 20, 1942.

# Five

# The Farny Years

George Simeon Weinberg was born in Rostov, Russia, on March 21, 1872. A well-educated man, he attended school in Russia, Belgium, France, and Germany before earning bachelor of science and master of science degrees at the Michigan School of Mining. His wife, Sylvia Farny Wurlitzer, was born in Ohio on September 21, 1869. George and Sylvia met when both were students in Berlin. George then immigrated to Sylvia's hometown of Cincinnati, Ohio, in 1895, and the two were married on February 19, 1896.

They had three children: Eugene Rudolph, born April 14, 1897; Cyril, born June 3, 1903; and Alice Leonie, born October 14, 1905. The family lived in St. Petersburg because George was an international consulting engineer with many clients in Russia. In 1915, with the Russian Revolution looming, they decided it was safer to return to America. They spent Christmas 1915 on the SS *Bergensford* and then settled near Sylvia's sister on Riverside Drive in New York City.

From there, they began a hunt for a sizable country estate. In 1917 they found the perfect one—Craftsman Farms—through a newspaper advertisement. On August 14, 1917, the Weinbergs purchased Gustav Stickley's 650 acres, his farm buildings, horses, cows, equipment, and even the furnishings of the log house.

America entered World War I that year, and George joined the engineering corps of the US Army as a commissioned officer. He completed his training that fall and sailed for France in December 1917, leaving Sylvia to handle the details of the sale and the management of the estate.

Following the war, anti-German public sentiment was strong. Despite his military service, there were concerns that his Germanic name would embarrass the children. As a result, in 1919, the Weinbergs formally assumed the surname of Sylvia's grandparents, Charles and Jeannette Farny.

Well educated, worldly, and financially secure, the story of the Farny years at Craftsman Farms is quite different than that of the Stickley years. However, both families shared a profound love for Craftsman Farms and the natural beauty to be found there. The Farnys and their descendants owned the property for more than 70 years.

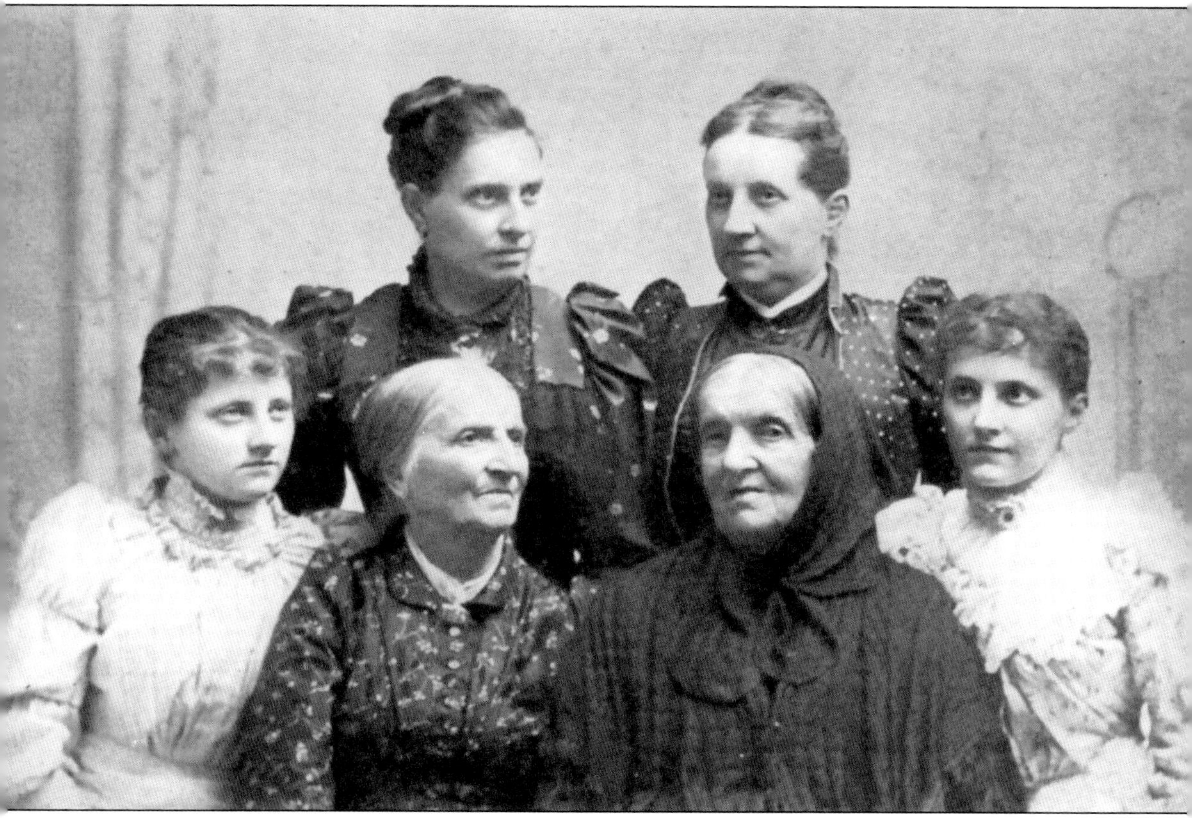

In 1853, at roughly the same time that Stickley's parents settled in Osceola, Wisconsin, a 22-year-old man named Franz Rudolph Wurlitzer emigrated from a German village in Saxony to seek his fortune in America. Shortly afterwards, a young girl named Leonie Farny and her family left the Alsace region in France to establish a new home in America. The two met a few years later in Cincinnati, Ohio, where Rudolph (as he was known) had established a successful business importing musical instruments. Rudolph Wurlitzer and Leonie Farny were married on September 19, 1868. Their first child, Sylvia Farny Wurlitzer, was born on September 21 of the following year. In this photograph from the late 1800s are three generations of women from the Wurlitzer Farny family. From left to right are (first row) Leonie Wurlitzer Eilers; Josephine Weingard; Leone's mother, Jeanette Farny; and Leonie Wurlitzer's daughter, Sylvia Farny Wurlitzer; (second row) Margaret Farny Strobel with her sister, Leonie Farny Wurlitzer. (Courtesy the author.)

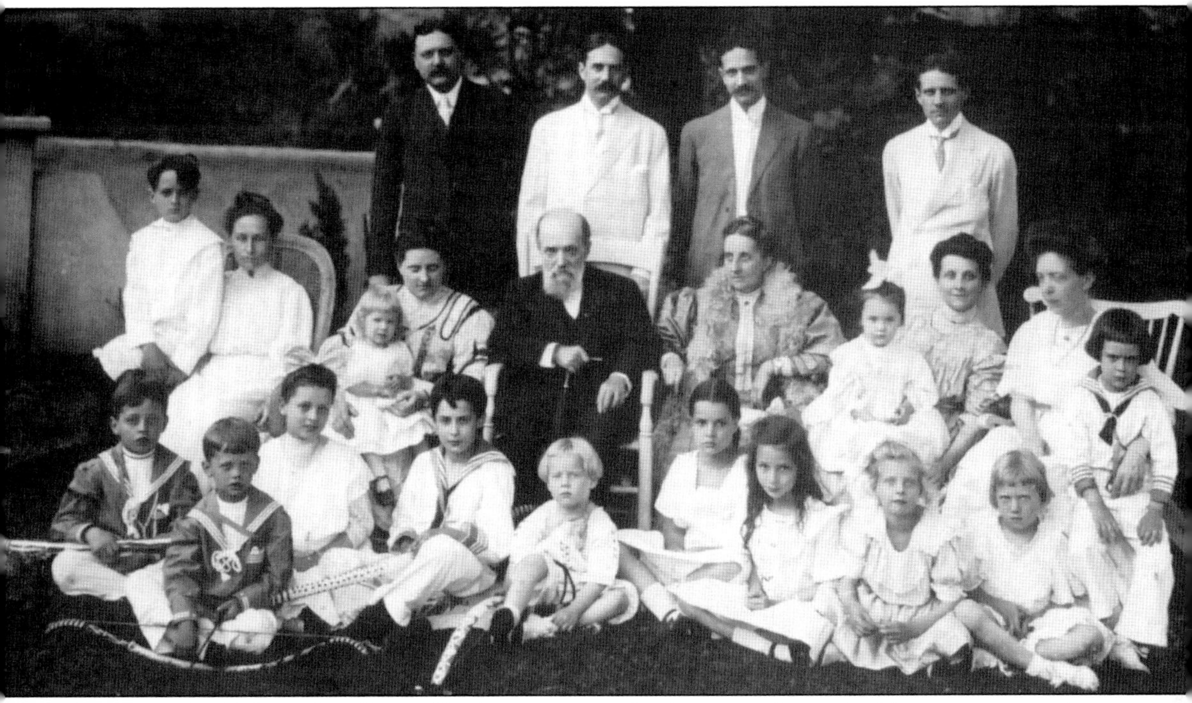

In the early 1890s, Sylvia Farny Wurlitzer went to Berlin to study music. Her parents sent her younger brother, Rudolph Henry, along to chaperone his elder sister. While in Berlin, Sylvia met George Weinberg, who was also a student in Berlin. When Rudolph reported that Sylvia was being courted, their father immediately insisted she return home. George was not to be deterred. He immigrated to Sylvia's hometown of Cincinnati, Ohio, in 1895. There he found employment as a draftsman and presented himself formally as a suitor. George Weinberg and Sylvia Farny Wurlitzer were married on February 19, 1896. This c. 1908 photograph was taken at the same time Stickley was buying land for Craftsman Farms. Sylvia is seated second from the right holding her daughter, Alice. Her son Cyril is to the right on the lap of Marie Richard Wurlitzer. Her son Eugene is seated in the front row fourth from the left. This original portrait, now lost, was owned by the late Natalie W. Griess and was given to the late Preston J. Kaufmann. (Courtesy Mechanical Music Press.)

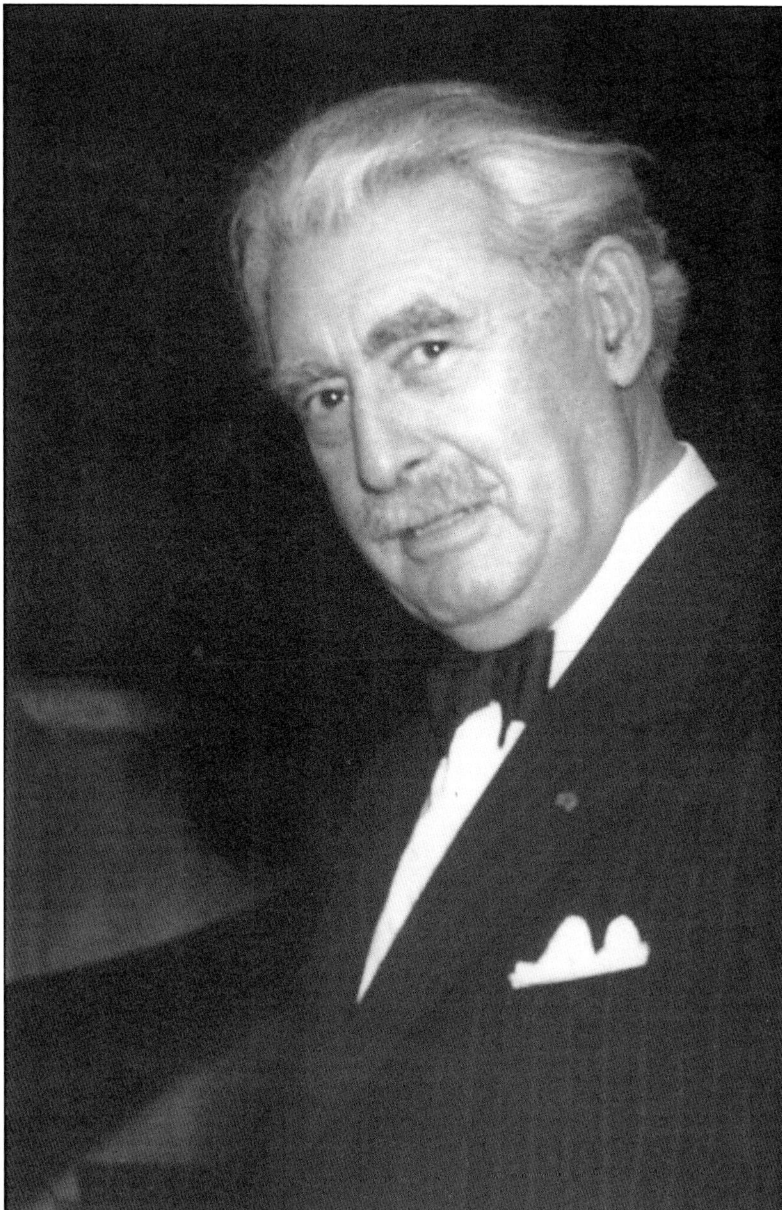

George continued his education in America, earning bachelor of science and master of science degrees at the Michigan School of Mining. After completing his military service in 1919, George became president of a whole wheat business, the Jireh Food Company, producing a special type of flour designed to meet the needs of diabetics. A civic-minded gentleman, he soon became very involved in his new life in New Jersey. He was a conservationist whose desire to preserve and protect the beautiful natural places in the nation was very much like that of Stickley. George served as chairman of the Morris County Planning Studies Commission and president of the New Jersey Parks and Recreation Association, and he was an active member of the New Jersey State Planning Commission as well as numerous other planning councils and associations. Today, a New Jersey state park in Rockaway Township is named for him. Shown here in an undated portrait, George was an international entrepreneur and a dynamic individual with a magnetic personality.

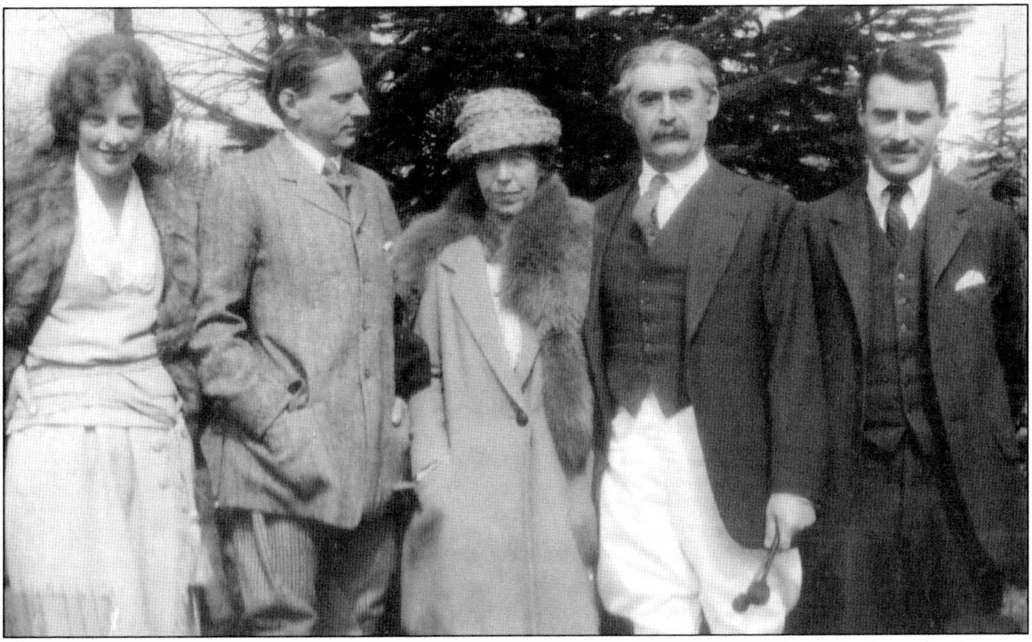

The 1920s photograph above shows two unidentified women and an unidentified man, Maj. George Farny, and son Eugene on the right. The undated photograph below shows Major Farny with grandson Gordon or Peter along with two dogs. Daughter-in-law Phyllis described her first meeting the major: "We were bowled over. He had on leather boots, white riding breeches, and he had a large moustache, bushy eyebrows and a tremendous head of hair. He had the blackest eyes and wonderfully ruddy brown cheeks. This exotic and glorious person seemed to like us too. It was amazing. What stories he could tell us, and then again, he could be gay and silly, and we would all feel exhilarated and witty ourselves. In no time we were calling him Uncle George, and eagerly looking forward to our next invitation to Craftsman Farms."

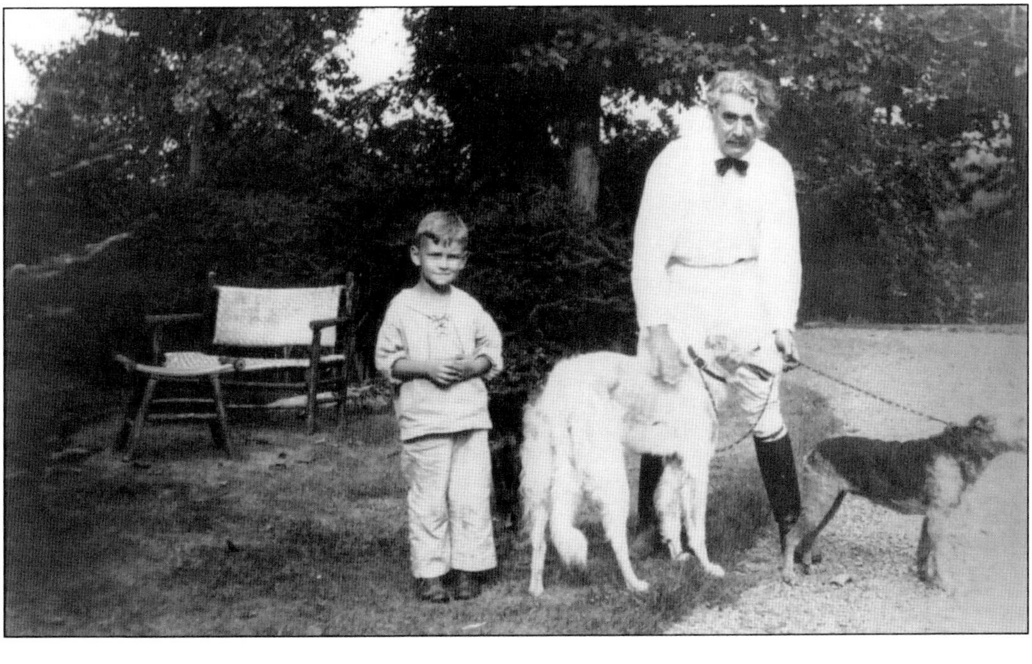

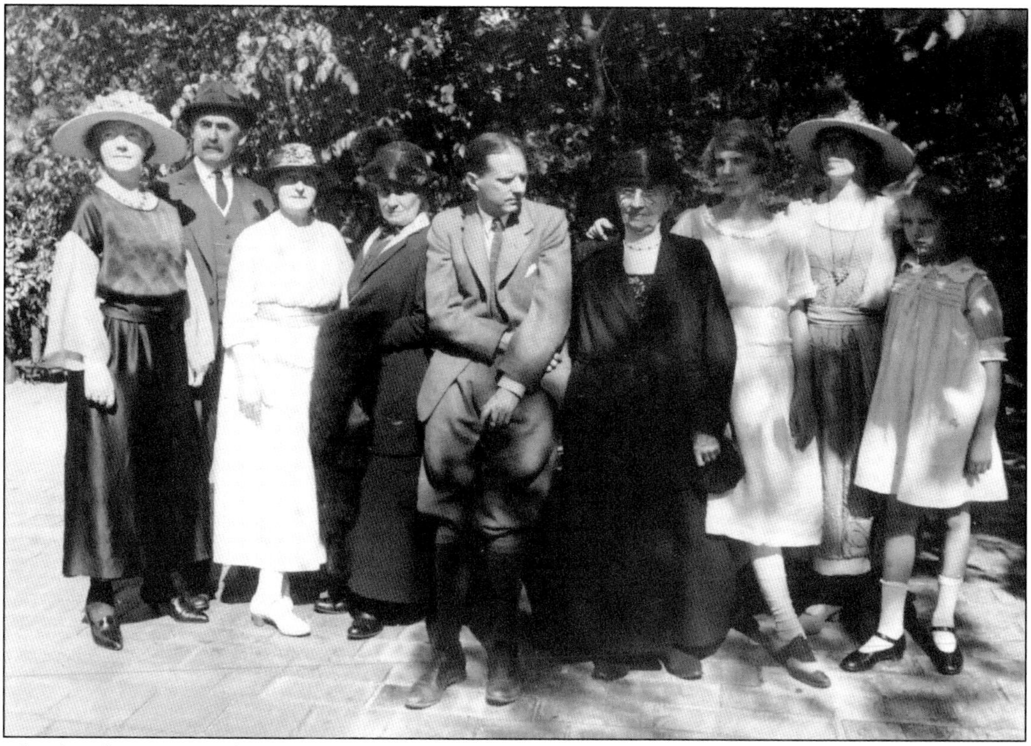

This family grouping, from the 1920s, shows Major Farny, second from the left, before his moustache had turned white. He is dressed splendidly in suit, vest, and hat. The petite woman in white standing beside him is Sylvia Farny. Her mother, Leonie Farny Wurlitzer, is believed to be the fourth from the right, dressed in black. The others are unidentified.

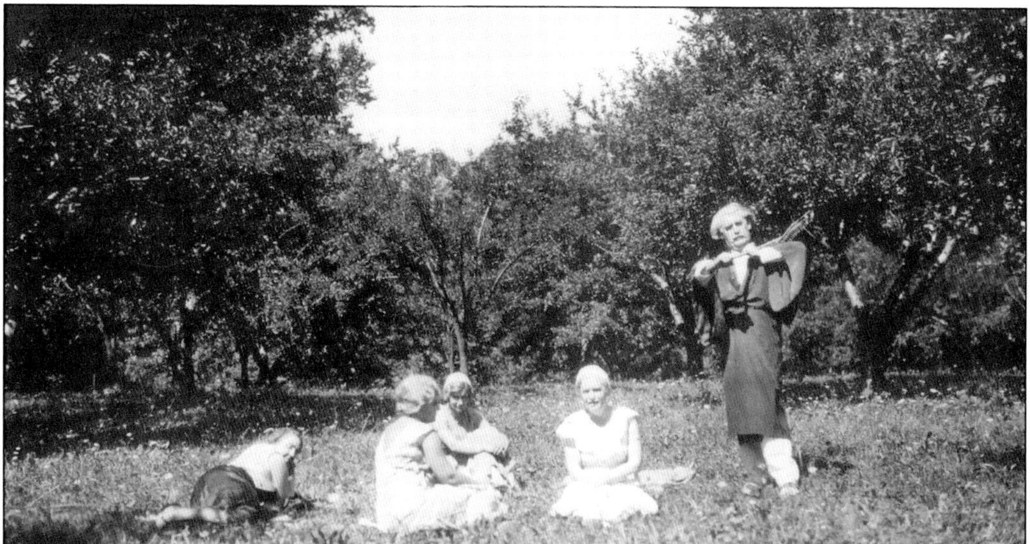

In this sunny day on the grounds of Craftsman Farms, Major Farny can be seen on the right, dressed in a kimono and acting out a scene near unidentified women. He enjoyed theater and directed at least one play for the Mountain Lakes Dramatic Guild. He was also a trustee of the American Institute of Music, Arts, and Drama.

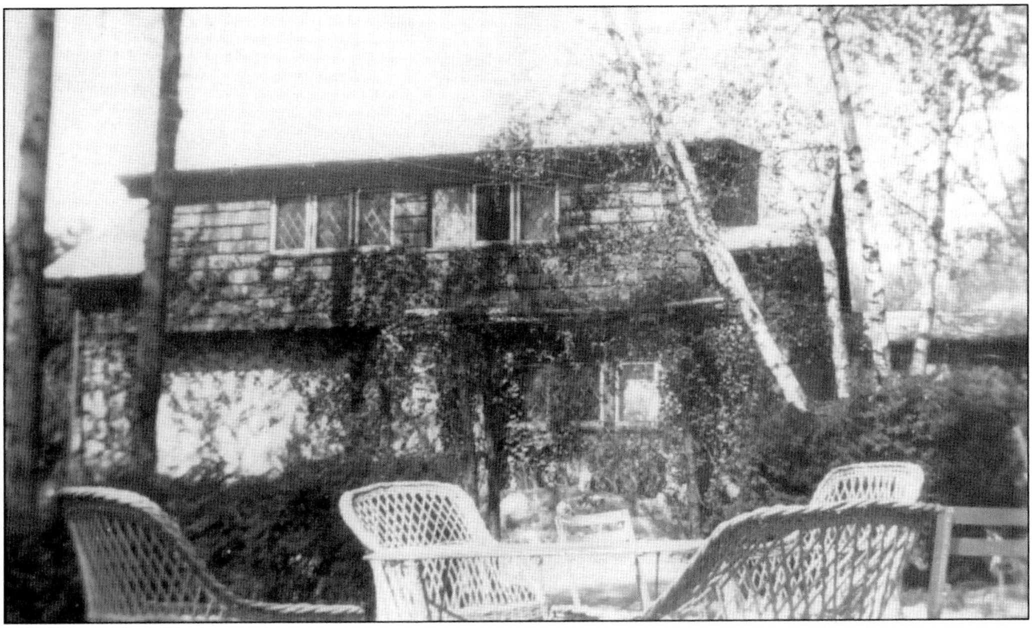

As soon as the Farny family settled into Craftsman Farms, the energetic Sylvia began adapting the property to their needs. She directed contractors to build the annex, which is shown here. It was needed to house the family's servants, who would have easy access to the main house via an attached and enclosed breezeway to the south side of the kitchen.

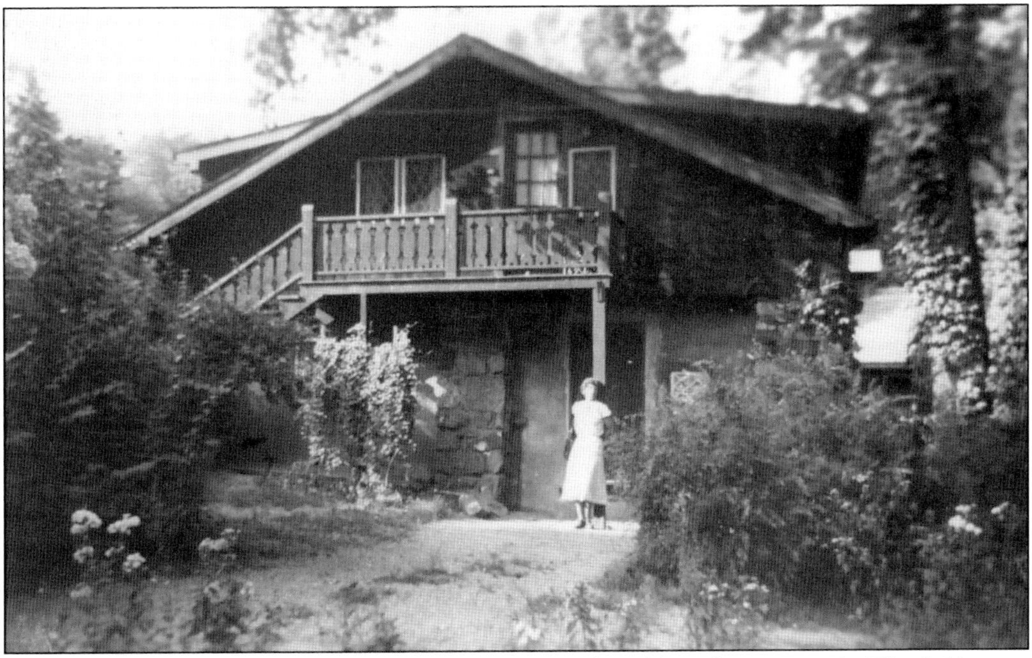

The annex was sensitively built within the stone piers of Stickley's dining pergola. The fieldstones of the original piers were left exposed. The exterior walls were sheathed in wood shingles, and diamond-paned casement windows were installed, all of which closely mimicked the Craftsman style of the earlier buildings. This view shows the exterior staircase later added to access a second-floor apartment.

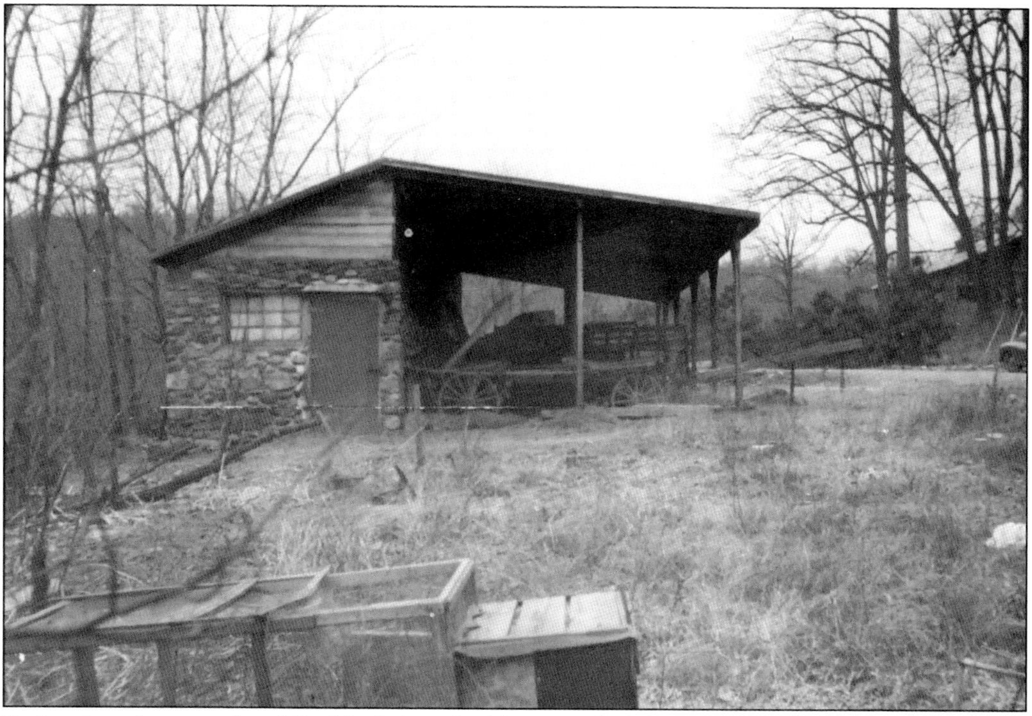

Only a few photographs of Stickley's farm buildings are known to exist. These two images show the side and front of the original calf barn, which was later converted into two apartments. For the first decade or so after the Farnys purchased Craftsman Farms, they actively farmed and operated the dairy, selling their products to the Morristown market. In the late 1920s, they gave up farming but rented the land to other farmers, which is very similar to the small farm cooperative that Stickley had envisioned for Craftsman Farms.

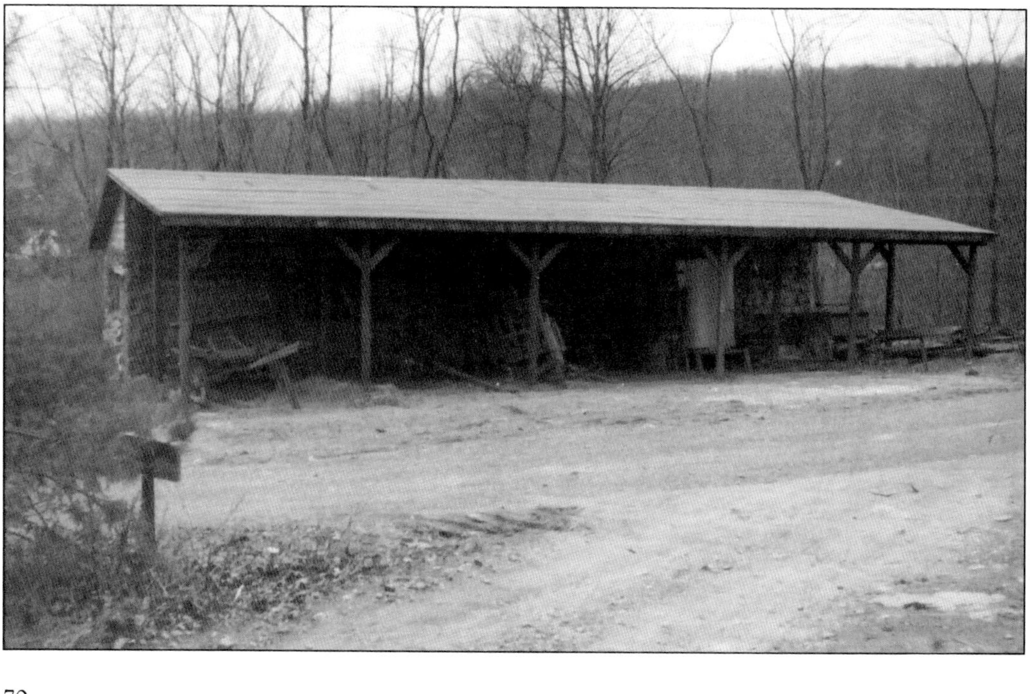

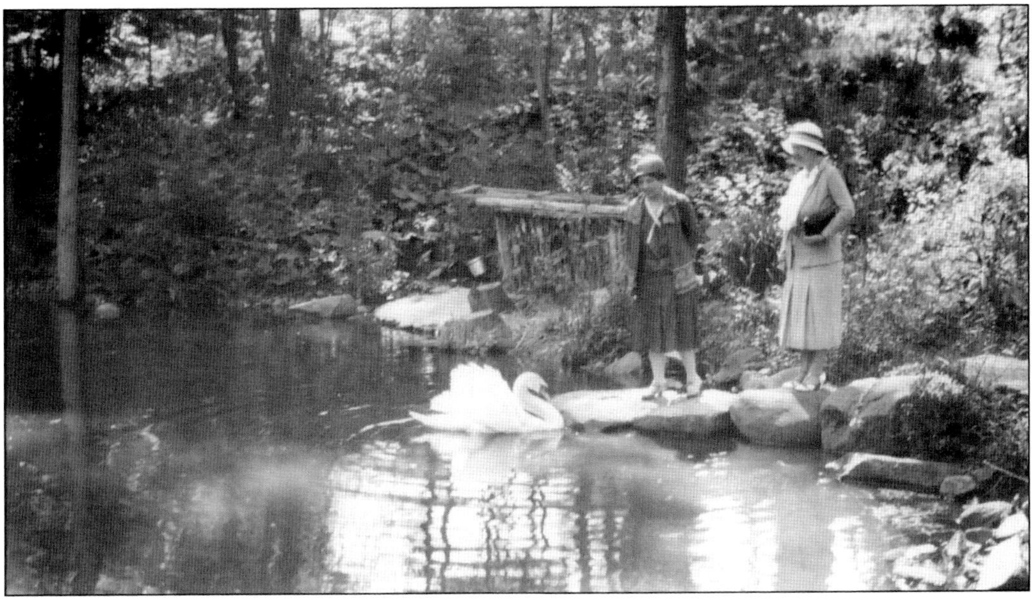

These photographs show the pond that Sylvia Farny added just south of the rose garden. Two unidentified ladies are admiring the swans. Sylvia was an avid gardener and was active in the local gardening club. Grandson Dave recollects, "She would be down on her hands and knees pulling weeds while Robert, her chauffeur, would stand nearby watching her work." Sylvia had grown up in a family known for its esthetic sensibilities, and the landscape flourished under her expertise. She was the niece of American artist William F. Farny, whose fine paintings of American Indians are in the collections of many museums, including the Smithsonian. Sylvia was also artistically talented and painted and sketched throughout her life.

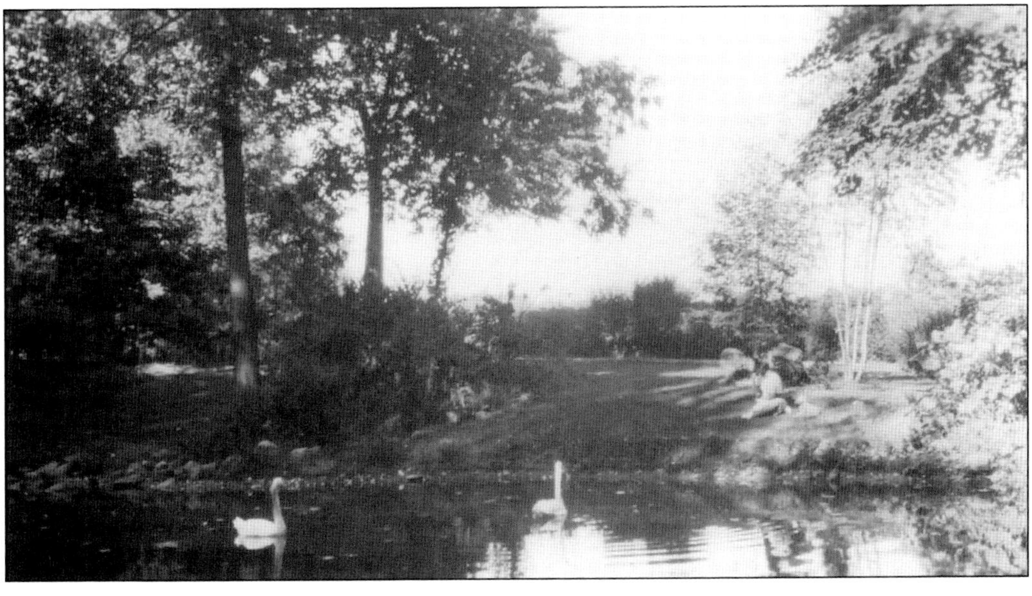

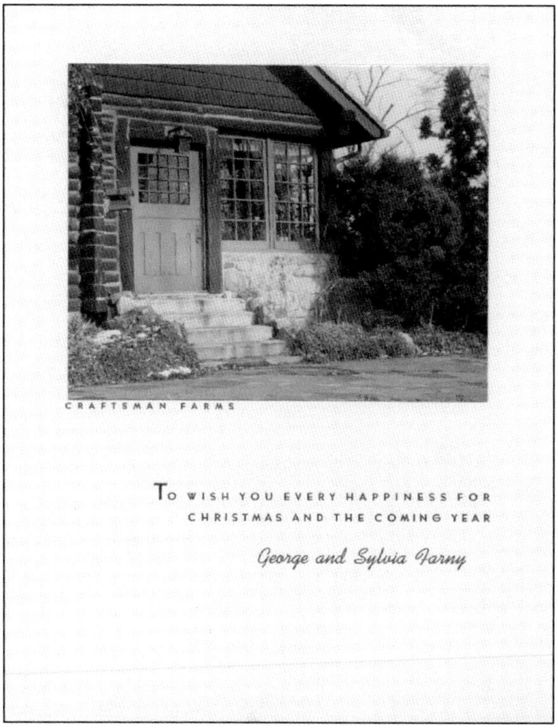


The pride the Farny family felt for their home is obvious in this beautiful 1930s Christmas card sent by George and Sylvia Farny. It shows the door from the terrace leading to the south end of the log house porch. Perennials are planted beside the steps, and masses of Sylvia's indoor plants can be seen through the window glass.

This signed Christmas card features one of the swans on the pond. The photograph may have been taken by Cyril Farny, who was an exceptional photographer. His interest was sparked by watching his mother, Sylvia, who was also a photographer, with a keen eye. She even developed her own black-and-white photographs.

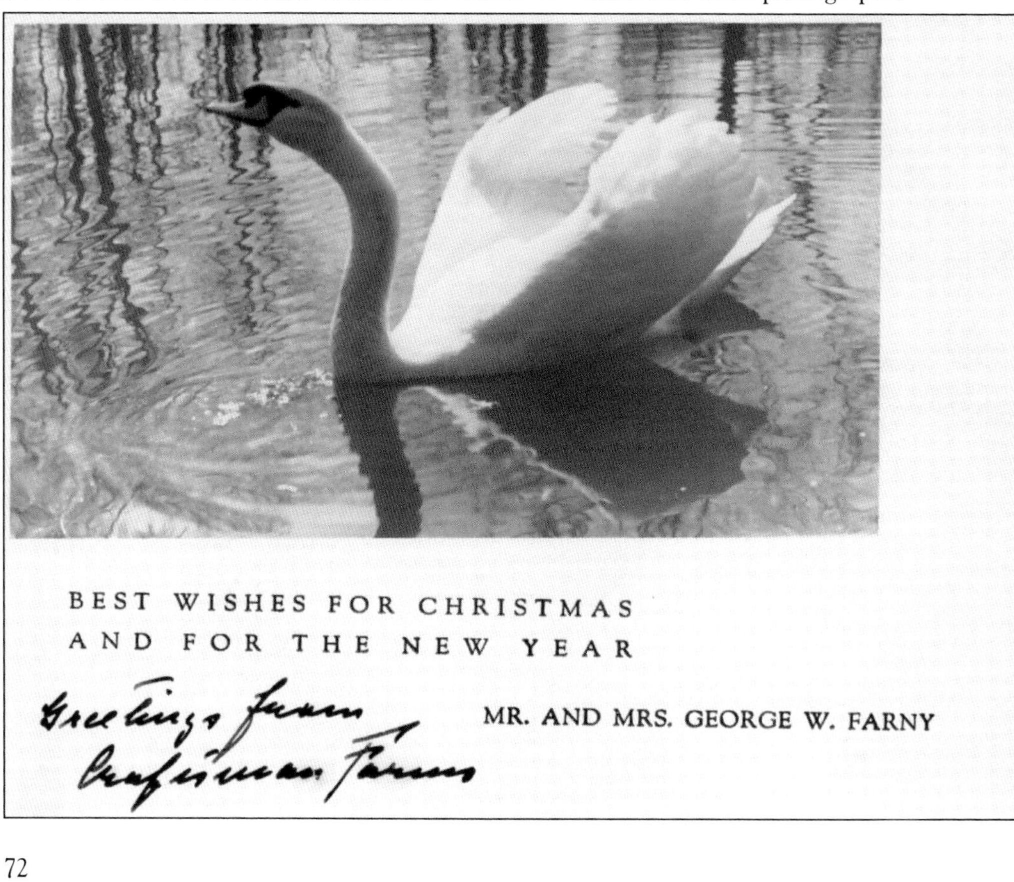

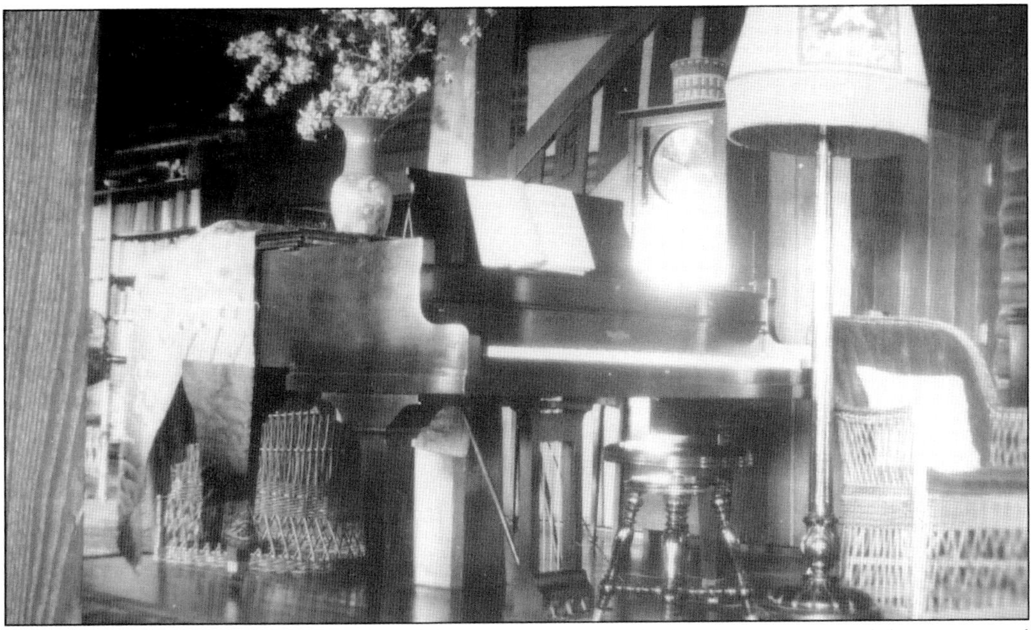

The centerpiece of this photograph of the living room in the log house is one of the two grand pianos Sylvia brought to Craftsman Farms. All the family members were musicians, and music was an important part of their family life for generations. Stickley's original furnishings, including the tall case clock and willow chairs, can still be seen in the room.

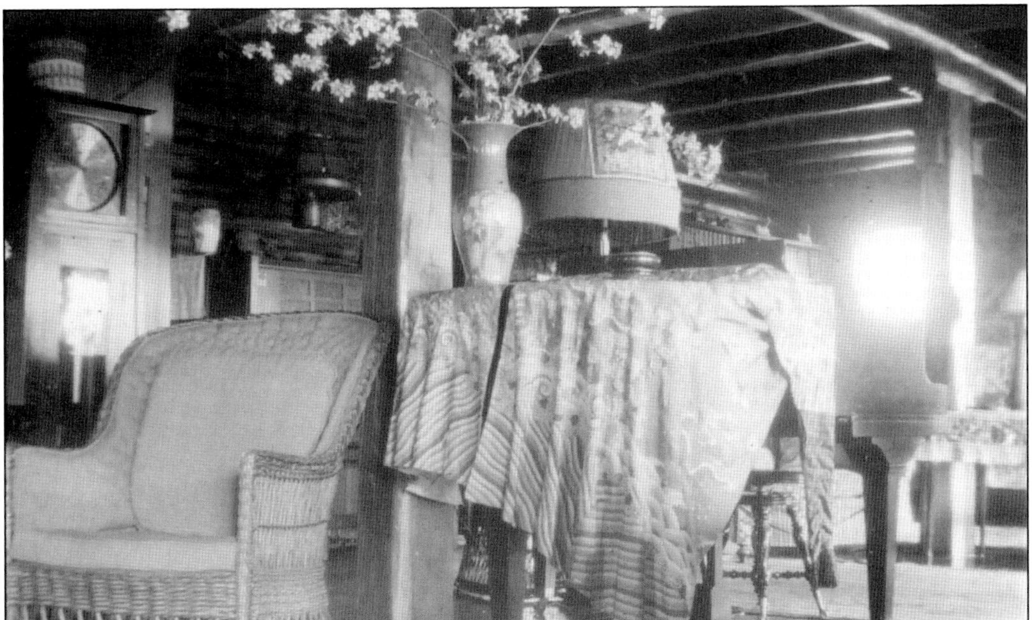

This photograph of the living room shows the blend of Stickley and Farny furnishings. A vase filled with blossoming branches, from Stickley's original orchard, sits atop the cloth-draped grand piano. In addition to two grand pianos, the rectangular form behind the willow chair is the top of Stickley's upright piano. Sylvia was known for presenting monthly musical evenings for neighbors.

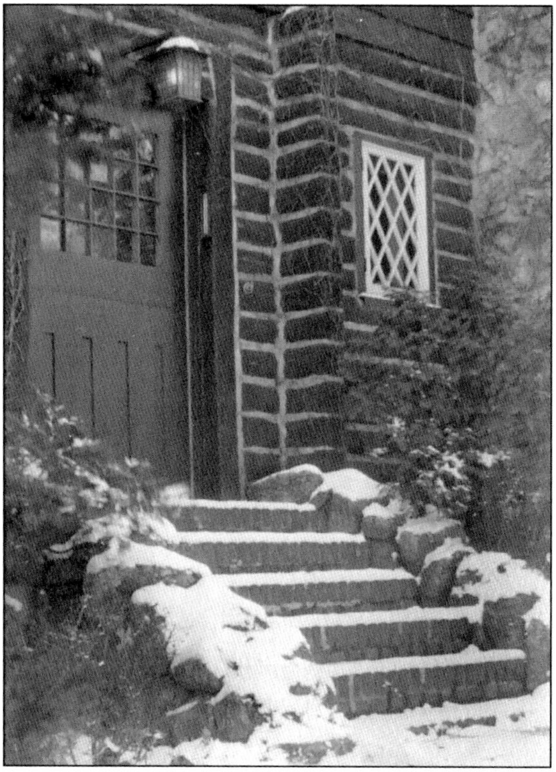

This 1930s image of the north entrance to the log house shows the corner of the house and its uncommon log construction method, known as "parallel log coursing." In the more standard "staggered log construction," logs are staggered like intertwined fingers. Instead, this building's logs are hewn flat on top and bottom, then notched to meet neatly at the corner.

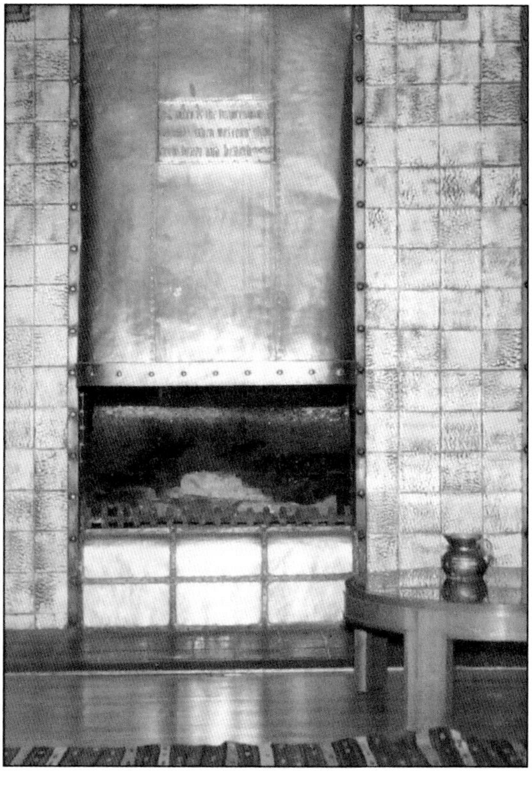

This interior view of a bedroom at the log house shows Farny furnishings near the Stickley fireplace. All the fireplaces at the log house have brass or copper hoods with mottoes hammered into them. This one expresses something both the Stickley and Farny families must have felt in the evening: "Sweet is the homecoming at eventide when welcome glows from heart and hearth."

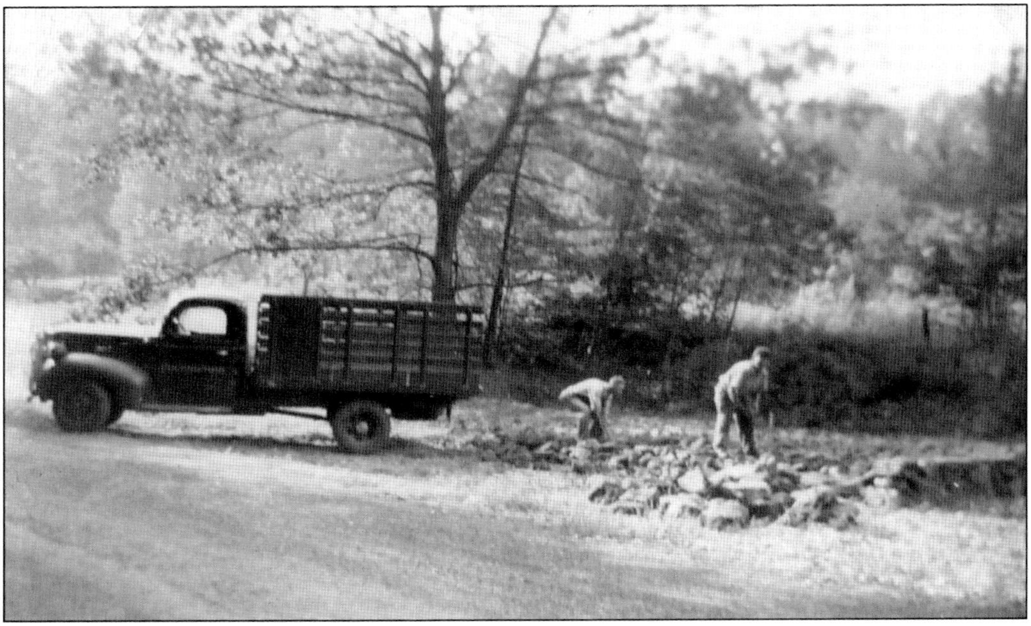

In this undated photograph, workers are repairing stonework at Craftsman Farms. When the Farnys moved into Craftsman Farms, Sylvia immediately began repairs and modifications to meet the needs of her family. In addition to creating the servants wing, she installed cement chinking between the logs, had a basement dug, and installed a furnace, all while the Major was away.

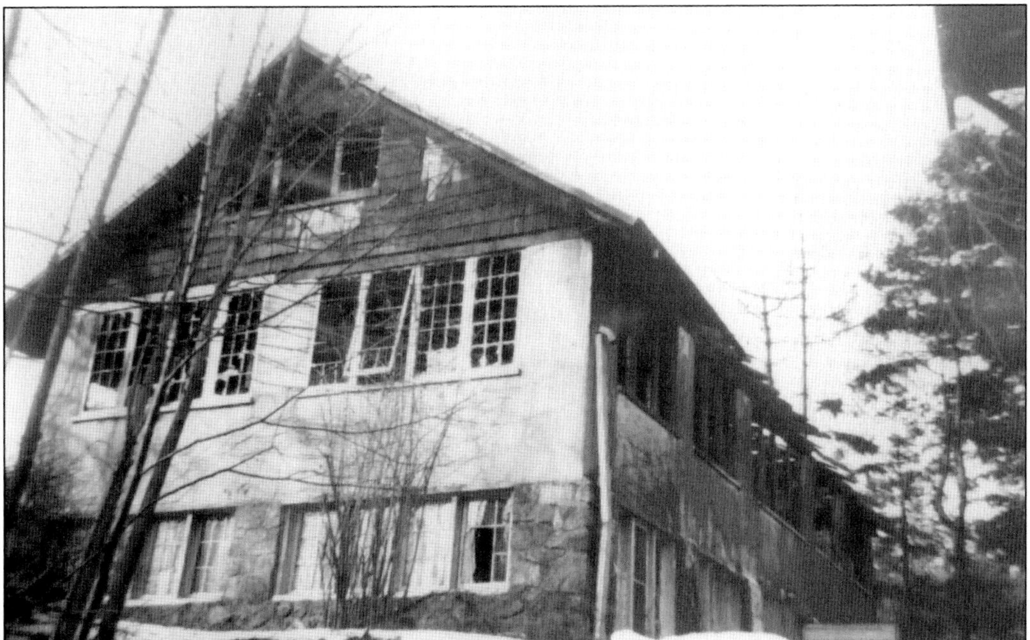

Very few images of this building are known to exist. It was built by Stickley and known as the garage and workshop. The major used it for his diabetic flour business for a while. This photograph was taken immediately after a fire in 1950, which forced the Farny family to raze the second floor and rebuild it on top of the original stone foundation.

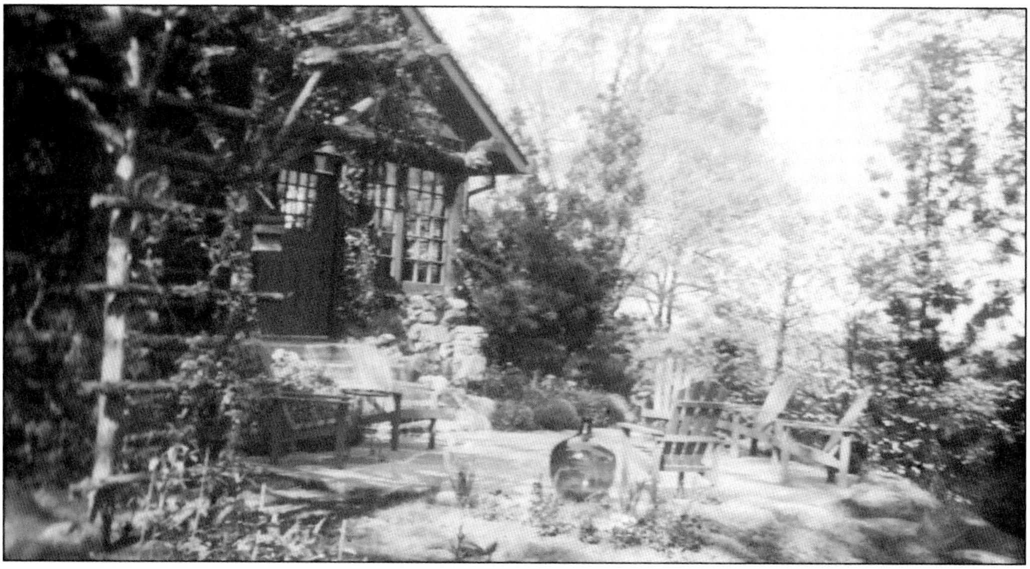

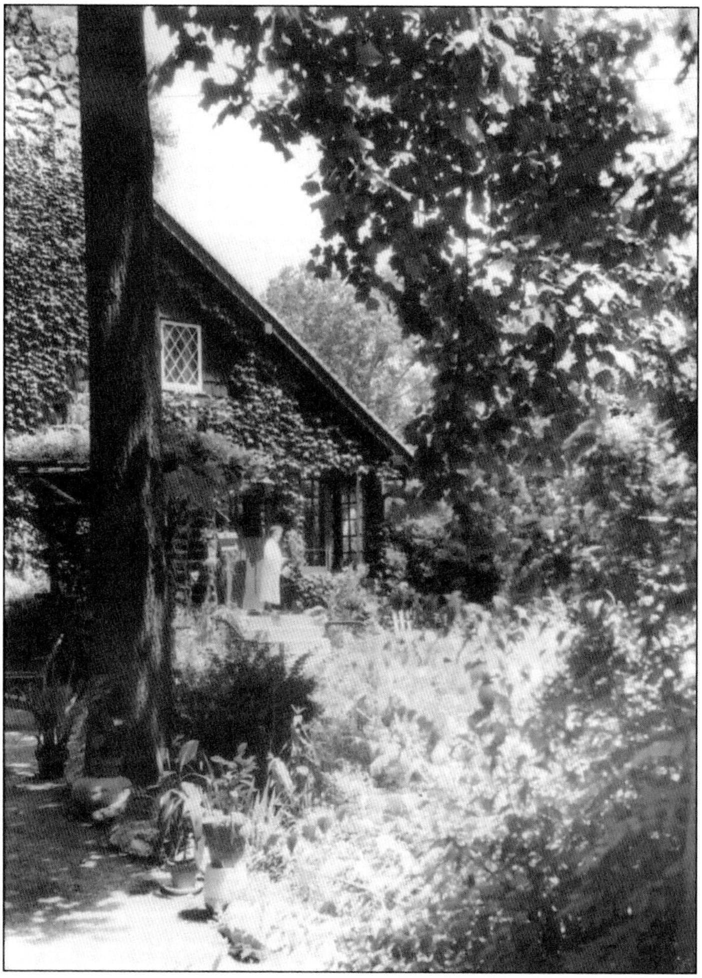

This view of the south side of the log house shows the terrace area with plenty of seating for the many friends and relatives who often visited Craftsman Farms. On the left of the photograph, a rustic arbor can be seen supporting vines. The semicircular slate patio is surrounded by perennials, shrubs, and trees, making it feel like an outdoor room.

The fairy-tale quality of the area to the south of the log house is best appreciated from this vantage point. The walkway on the left is lined with numerous potted plants set against the backdrop of stone edging, ferns, and other woodland plants. Sylvia Farny can be seen dressed in white near the house as she tends to her beloved garden.

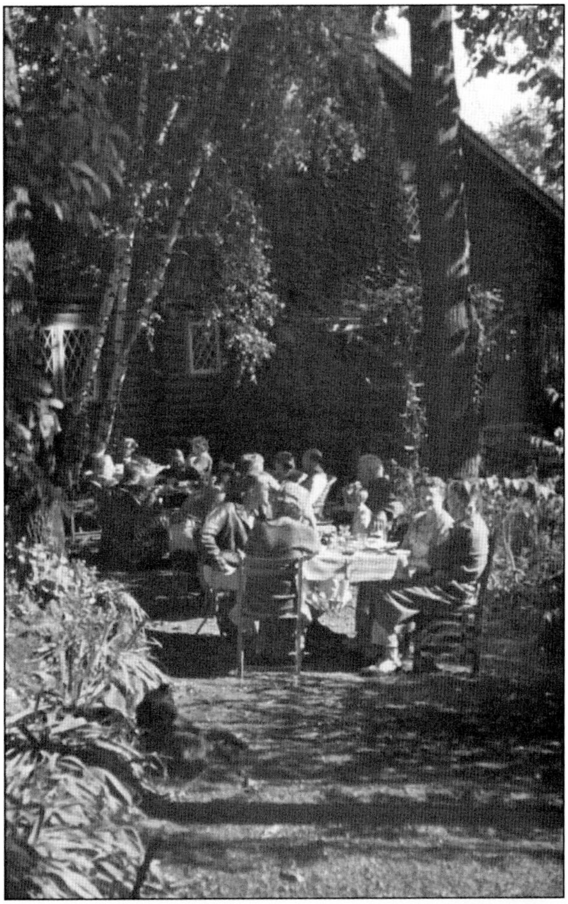

These two scenes of dinners at Craftsman Farms are a small glimpse of the entertaining that took place at Craftsman Farms. Daughter-in-law Phyllis recalled, "Once a month, Sylvia and George had a dinner party for perhaps 30 people at Craftsman Farms. People sang or played various instruments and George accompanied them all on the piano. Sunday mornings, friends of theirs from the Oranges and Short Hills came up to talk, walk, play medicine ball, and have brunch. One couple didn't miss a single Sunday for seven years. . . . Eugene and Margaret and their three girls moved to the Farms. Alice and Dick and their two children visited a great deal." Relatives and friends frequently stayed in the cottages and, as in Stickley's time, they ate all their meals in the log house dining room.

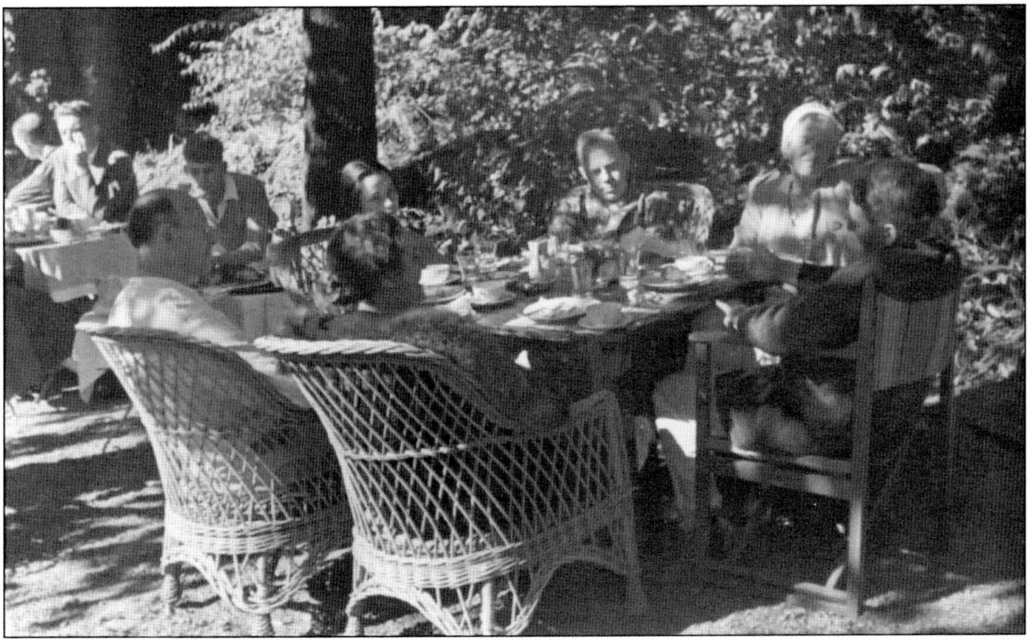

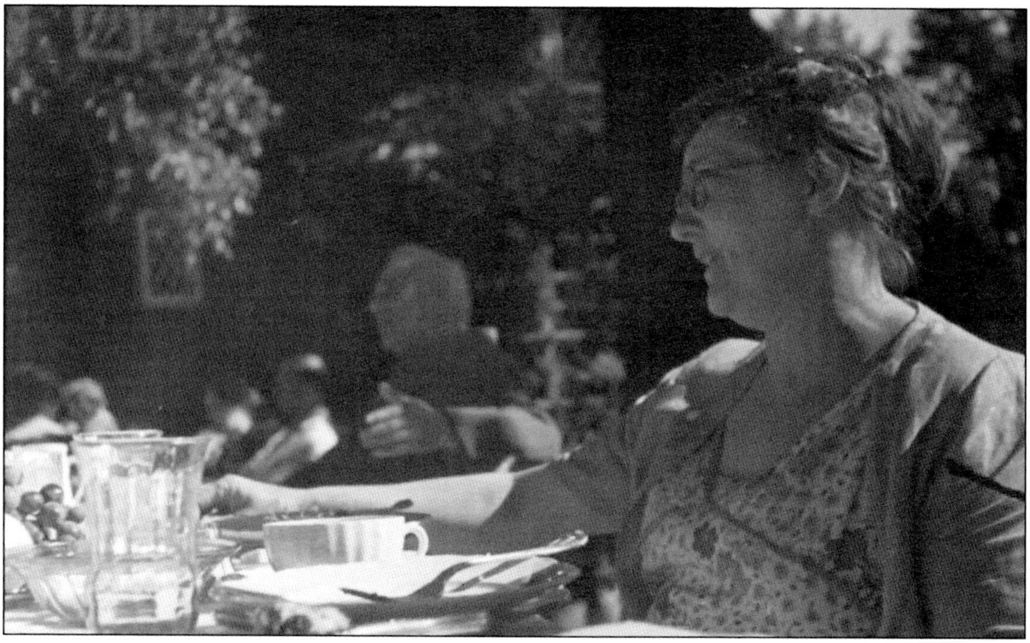

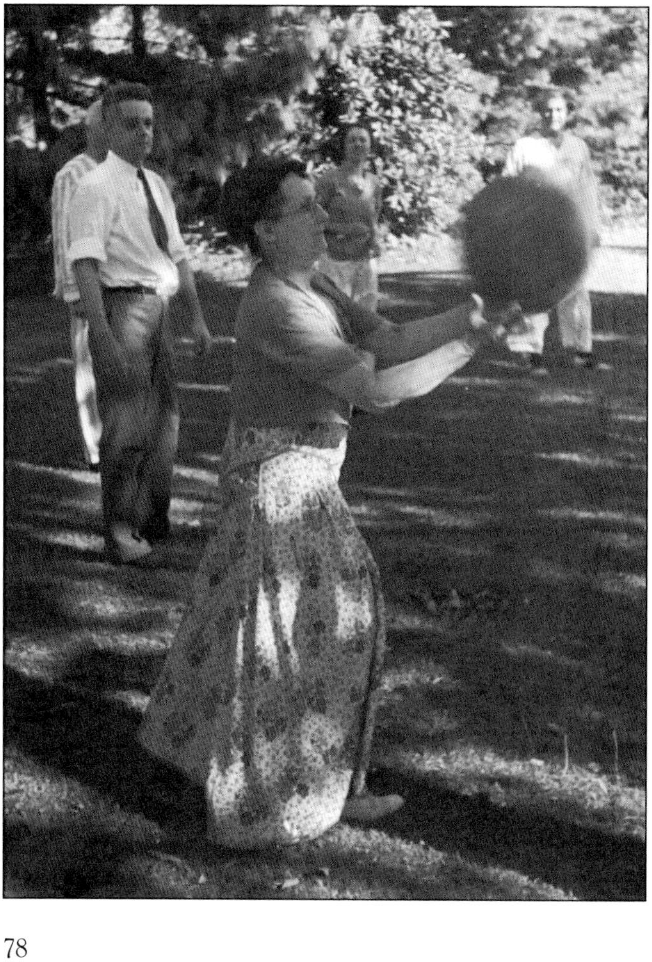

Sylvia Farny is relaxing at one of the many outdoor parties that took place at Craftsman Farms. She was known to be a marvelous hostess. Daughter-in-law Phyllis described her, saying Farny had "the brightest green eyes and the most hospitable manner." She continued, "and she talked about things she assumed we would be interested in—plays, book, politics."

Though quite petite, Sylvia was an active and lively woman who thoroughly enjoyed outdoor activities at Craftsman Farms. She can be seen in this 1920s photograph playing the family favorite game of medicine ball. Directly behind the ball, dressed in white, is Major Farny.

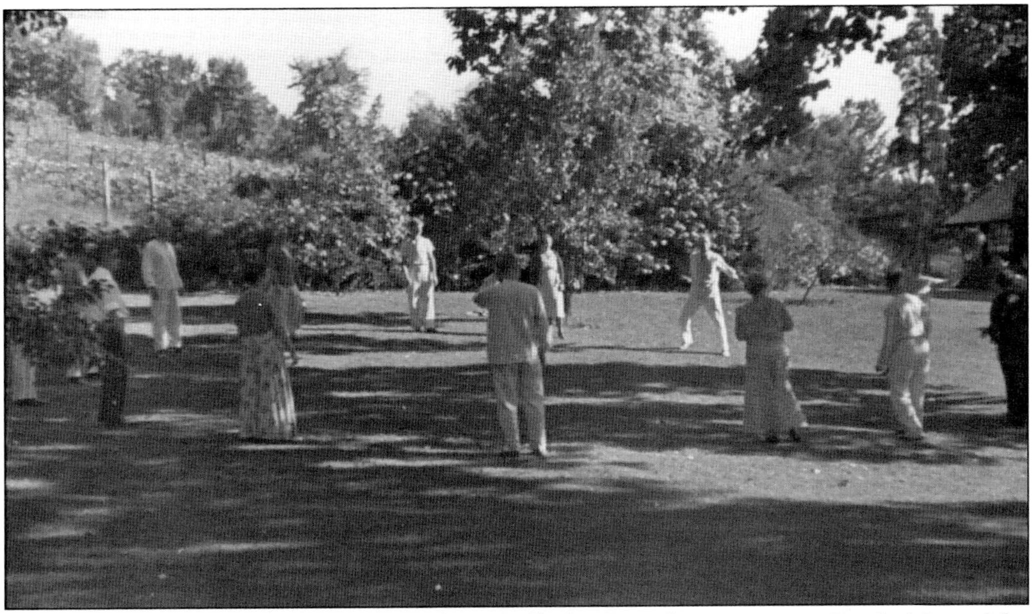

Family and friends joined in playing medicine ball after brunch on many weekends. Most of the men are dressed in loose white clothing, while the women play in their long flowing skirts. The hillside behind them on the left is the terraced vineyard still thriving from Stickley's days. The roof of the north cottage is visible on the far right.

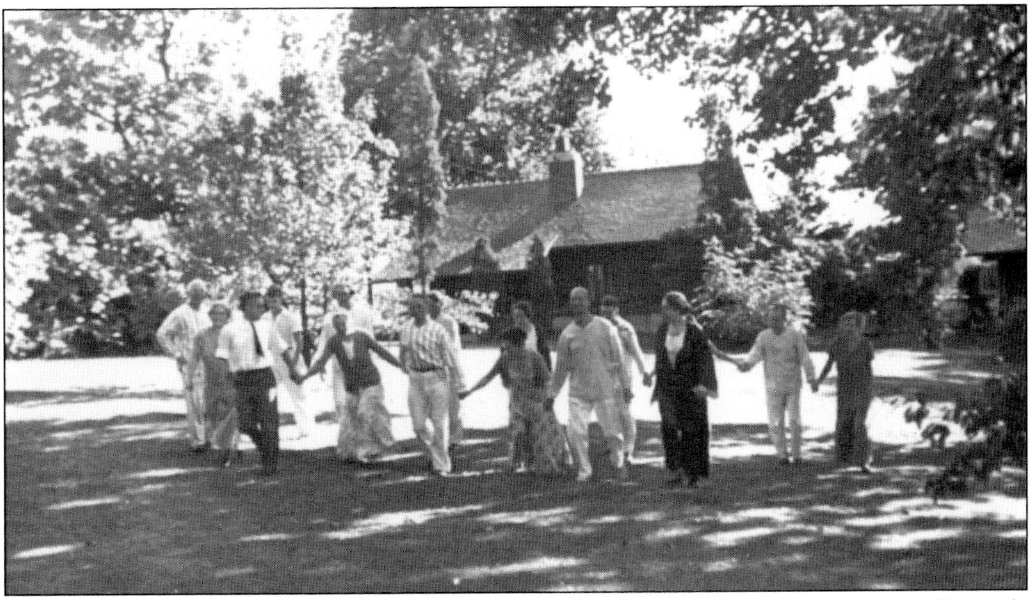

Guests in this Sunday brunch group take hold of each other's hands as they energetically stride across the lawn to the west of the cottages in preparation for lively lawn games. The small woman in the front center, seen holding hands with a man in a striped shirt, is their hostess, Sylvia Farny.

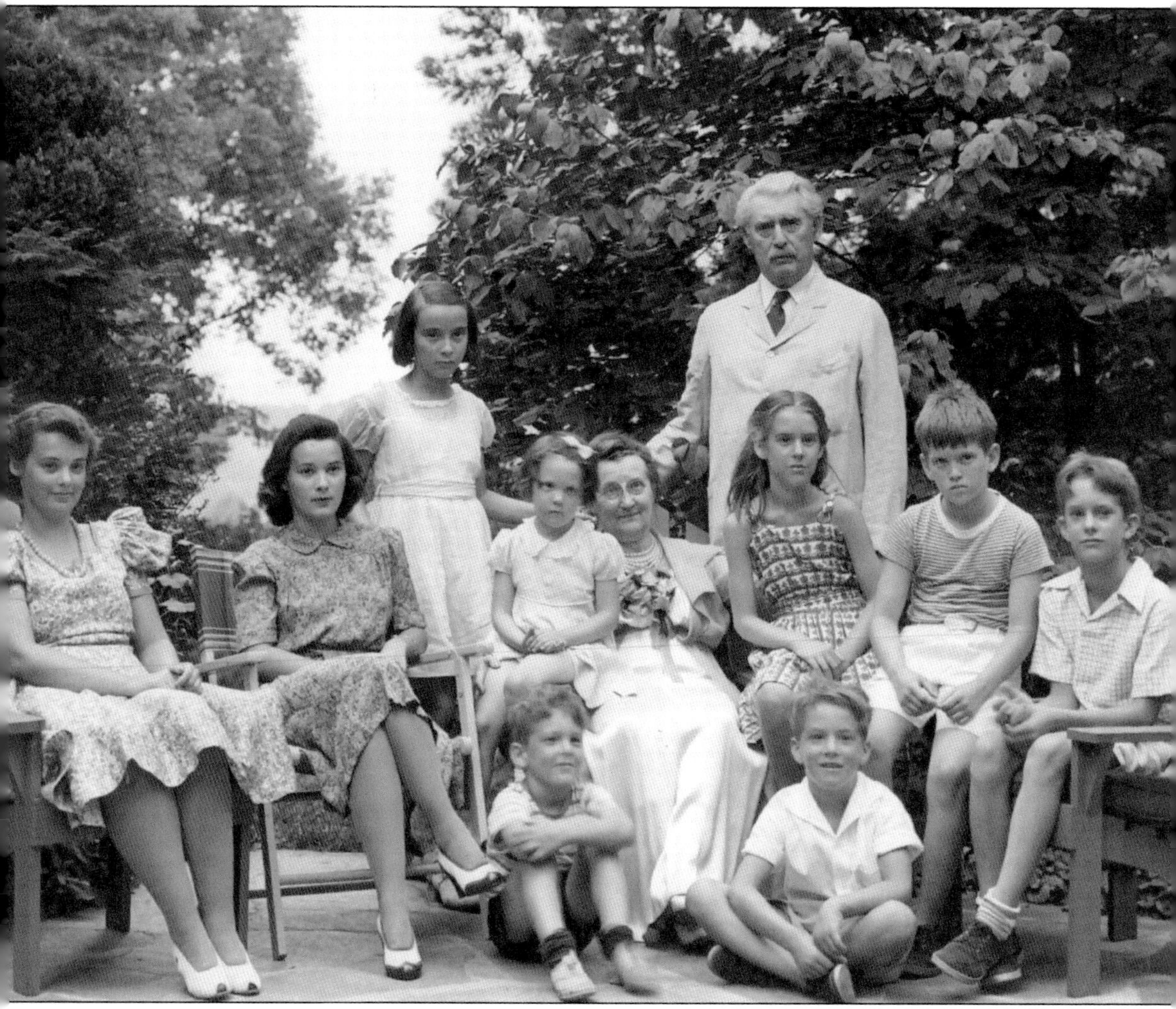

Sylvia and George Farny are surrounded by nine grandchildren in the late 1930s in this photograph, taken on the slate terrace during one of the family's garden parties. Standing in back are Eugene's daughter Sylvia "Sibby" Farny and Major Farny. Seated from the left to right are Eugene's daughters Sally and Josie Farny; Alice's daughter Patsy Wood; Sylvia Farny; Cyril's daughter Evelyn "Evvie" Farny; Alice's son Peter Wood; and Cyril's son Christopher "Kit" Farny. Cyril's sons Michael (left) and David Farny (right) are in front. The grandchildren had many memories of "Grandma Farmsz," as they called Sylvia. She taught them to speak French, though the boys would often prefer running around outdoors to having French lessons. Granddaughter Sally remembers the gentle way her grandmother corrected them when teaching them to play piano. (Courtesy the Stickley Museum at Craftsman Farms, David Farny Collection.)

In addition to the regular weekend gatherings and the overnight guests, the Farnys hosted many large parties and concerts on the property. This formal event is believed to have taken place in June 1938 or 1939. Sylvia Farny, who would have been about 70 years old, is seen here greeting one of the guests. (Courtesy the Stickley Museum at Craftsman Farms, David Farny Collection.)

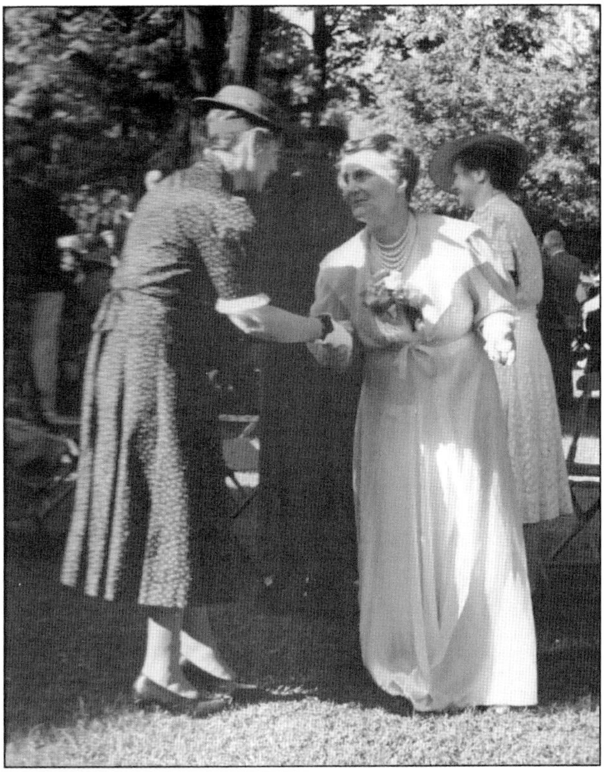

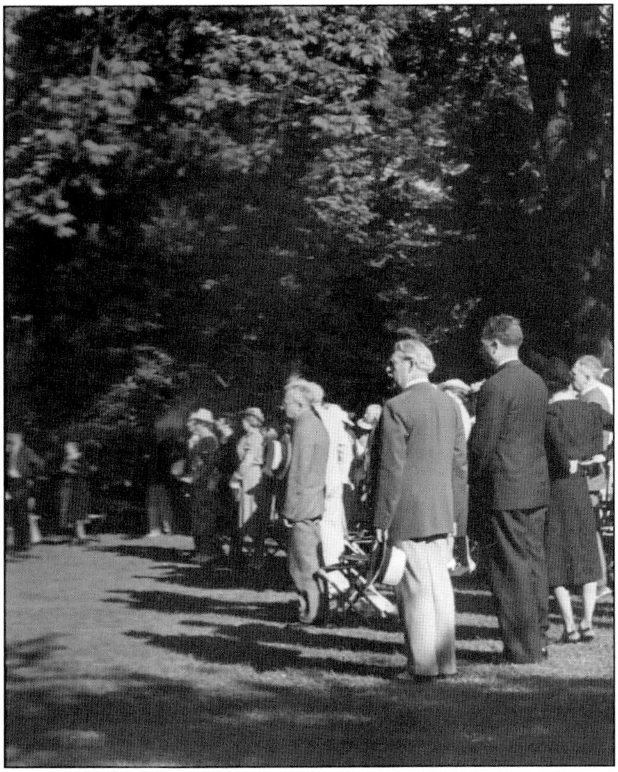

Music played a big role in the lives of Sylvia and George Farny, and this summer party was no exception. In this photograph, guests are standing for the national anthem, which was performed by the Morristown Junior Bugle and Drum Corps. Major Farny can be seen closest to the camera, holding his hat in his hand. (Courtesy the Stickley Museum at Craftsman Farms, David Farny Collection.)

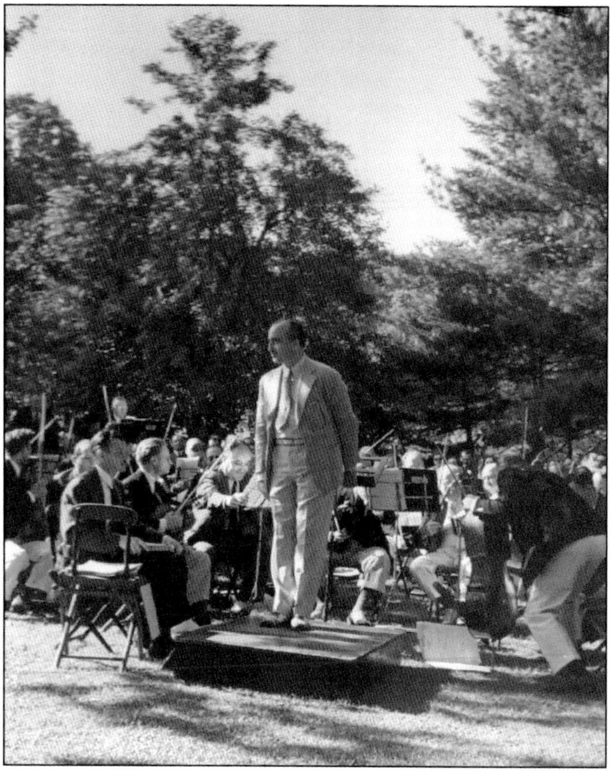

The Farnys' garden parties often included performances. Family lore says that they even had the New York Philharmonic Orchestra perform. Papers from the concert shown here indicate this event was presented by the Works Progress Administration's Federal Music Project, and the national director, Dr. Nicholas Sokoloff, is acknowledged. (Courtesy the Stickley Museum at Craftsman Farms, David Farny Collection.)

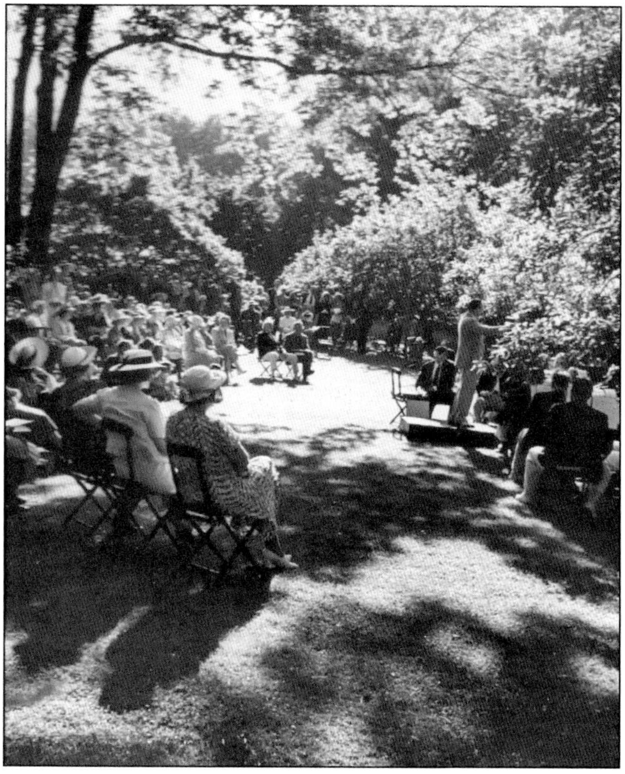

This image shows the guests seated in a semicircle on the lawn enjoying the concert. The women are in stylish dresses and hats, and the men are dressed with jackets and ties. Existing paperwork shows that the Federal Opera Company of New Jersey performed selections form Bizet's *Carmen* and Verdi's *Rigoletto*, *La Traviata*, and *Aida*, among other works. (Courtesy the Stickley Museum at Craftsman Farms, David Farny Collection.)

The garden party concert is an example of the intellectually stimulating, cultured, and worldly life the Farnys led at Craftsman Farms. Family members in this close-up of guests enjoying the garden party concert may include Eugene's daughter, Josie, wearing the broad-brimmed white hat. Alice Farny Wood, the daughter of Sylvia and George, may be the woman smiling at the camera. (Courtesy the Stickley Museum at Craftsman Farms, David Farny Collection.)

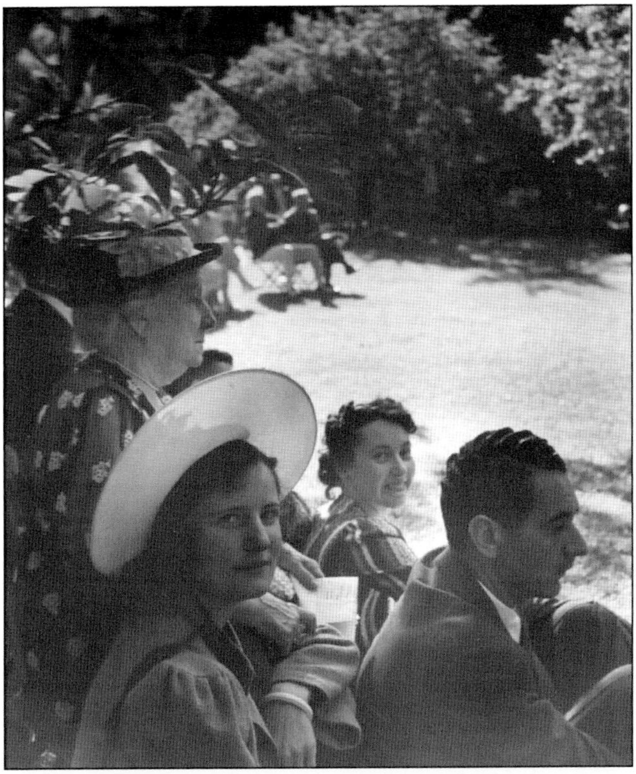

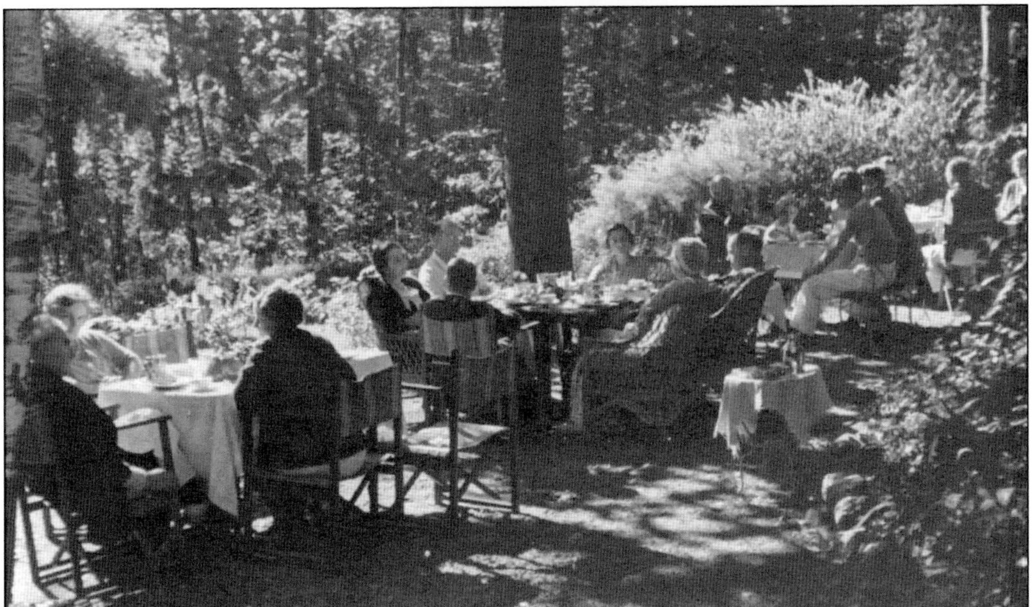

Stickley willow furniture mixes with newer furnishings in this afternoon meal on the grounds at Craftsman Farms. The Farnys had left all their furniture in Russia, but the log house had come fully furnished. They could have purchased new furnishings but recognized how perfectly the Stickley furniture harmonized with the house and their own appreciation for this rustic country retreat. (Courtesy the Stickley Museum at Craftsman Farms, David Farny Collection.)

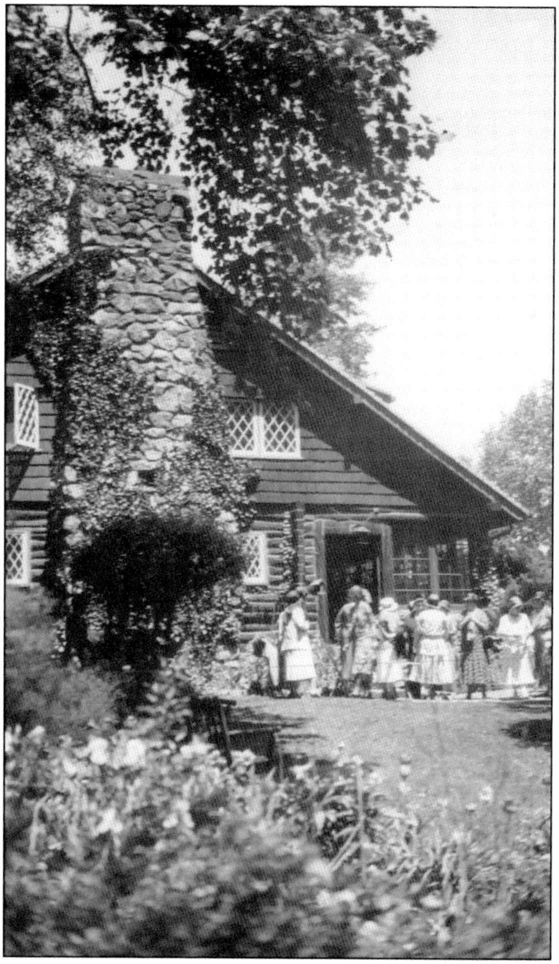

A group of stylishly dressed ladies gathers at the south side of the log house. This photograph shows that, more than 20 years after Gustav Stickley and his family left Craftsman Farms, the gardens still flourished with blooms of all sorts—just as Stickley had wanted for what he called his "Garden of Eden." (Courtesy the Stickley Museum at Craftsman Farms, David Farny Collection.)

The ladies are gathered to begin a tour of the house. While this grouping is from a garden party, Sylvia Farny would have repeated this scene many times. She was a member of a church and a garden club, as well as an officer of the Women's Auxiliary of the American Society of Mechanical Engineers. (Courtesy the Stickley Museum at Craftsman Farms, David Farny Collection.)

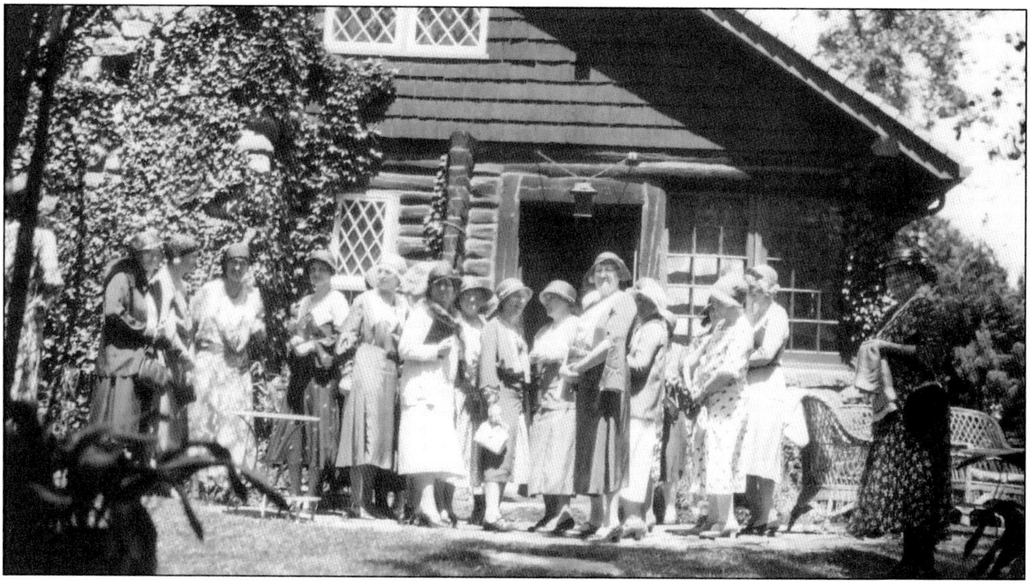

Always a commanding presence, George was known as "the major" for most of his life. Elegantly dressed, Sylvia and George Farny are enjoying a quiet moment together at Craftsman Farms. The photograph is undated, but Sylvia is wearing the gown she wore at the garden party believed to take place in 1938 or 1939.

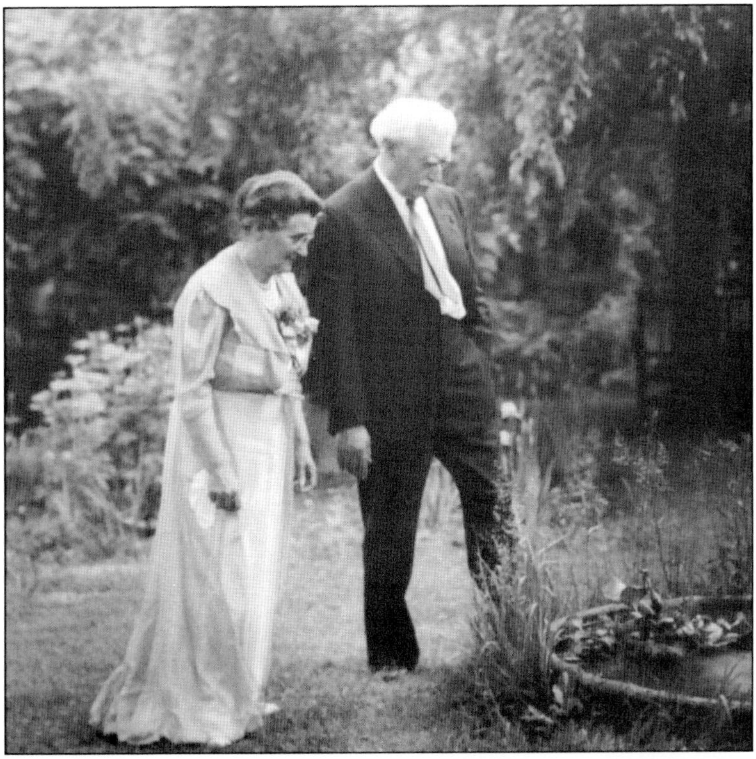

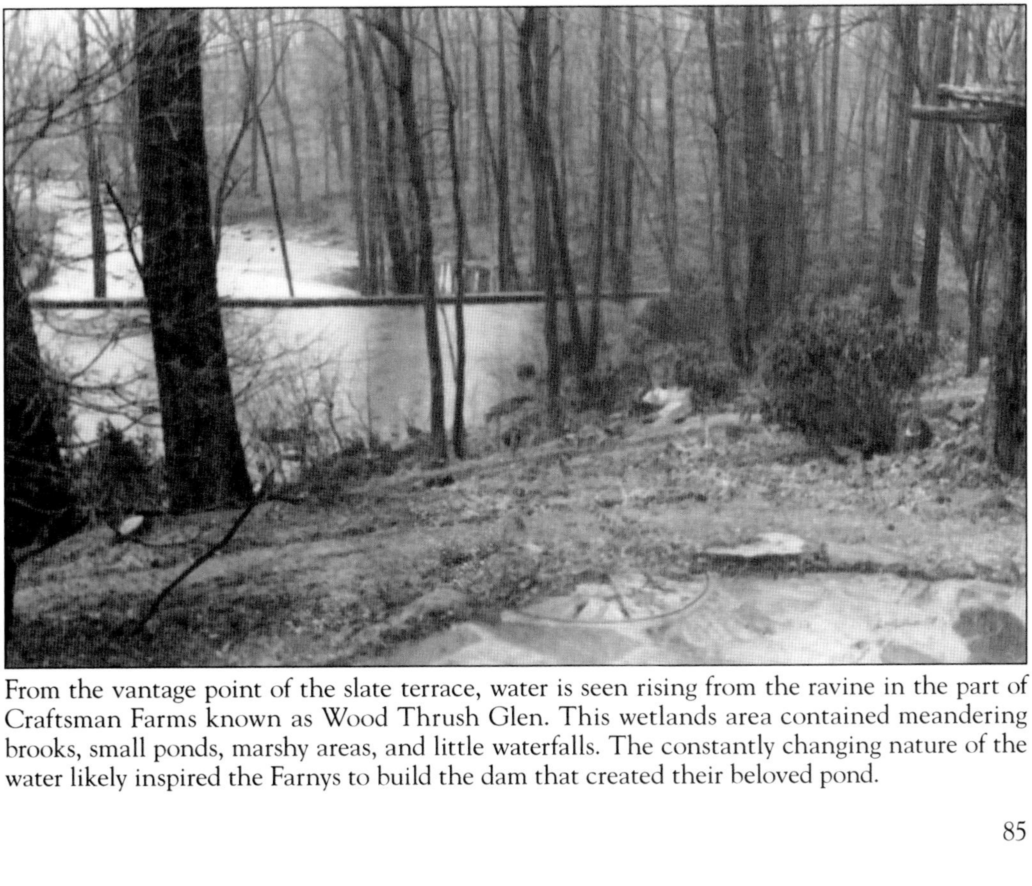

From the vantage point of the slate terrace, water is seen rising from the ravine in the part of Craftsman Farms known as Wood Thrush Glen. This wetlands area contained meandering brooks, small ponds, marshy areas, and little waterfalls. The constantly changing nature of the water likely inspired the Farnys to build the dam that created their beloved pond.

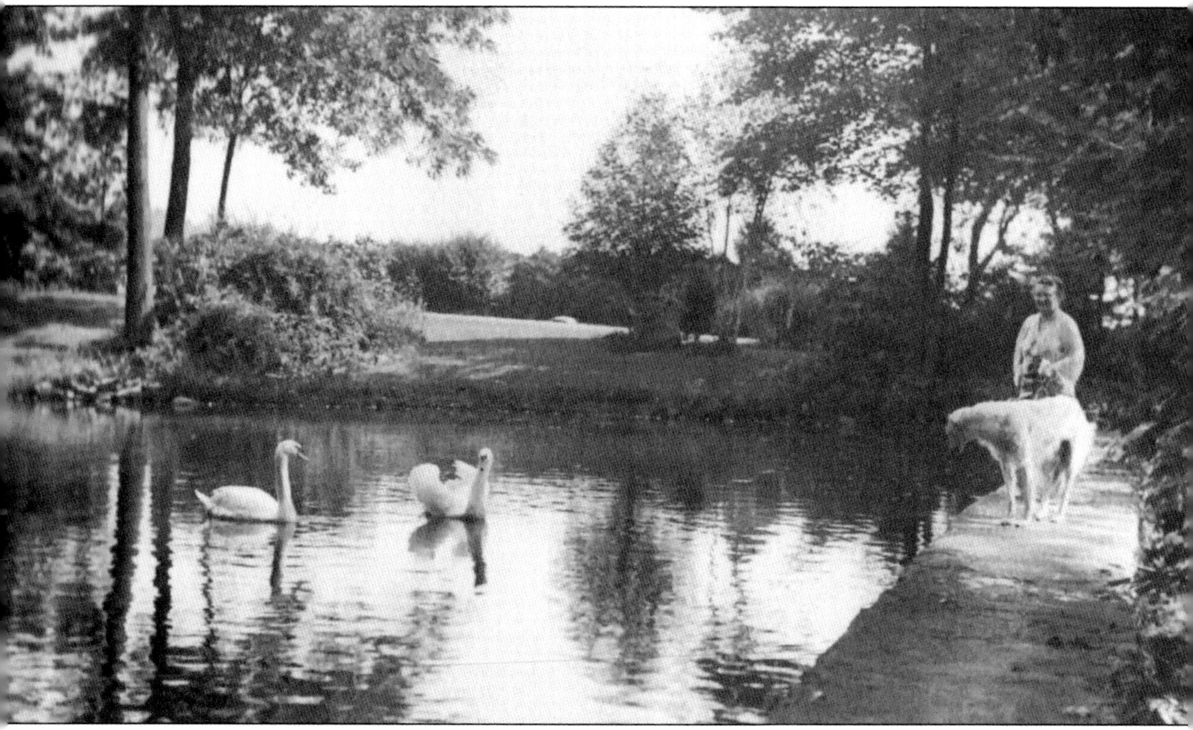

Sylvia Farny is standing in this serene and picturesque landscape that bears the stamp of her esthetic sensibilities. Two of her swans float gracefully on the pond that she created just south of the rose garden at Craftsman Farms. One of her husband's Russian wolfhounds is at her side. Both Sylvia and George loved this breed, and for some time the family had a busy kennel on the property. George died on September 1, 1941, at the age of 69. Later that year, Sylvia transferred the ownership of the property to her three children and their spouses. She continued to live on the property and enjoy sights such as this one until her death on March 10, 1952, at the age of 82.

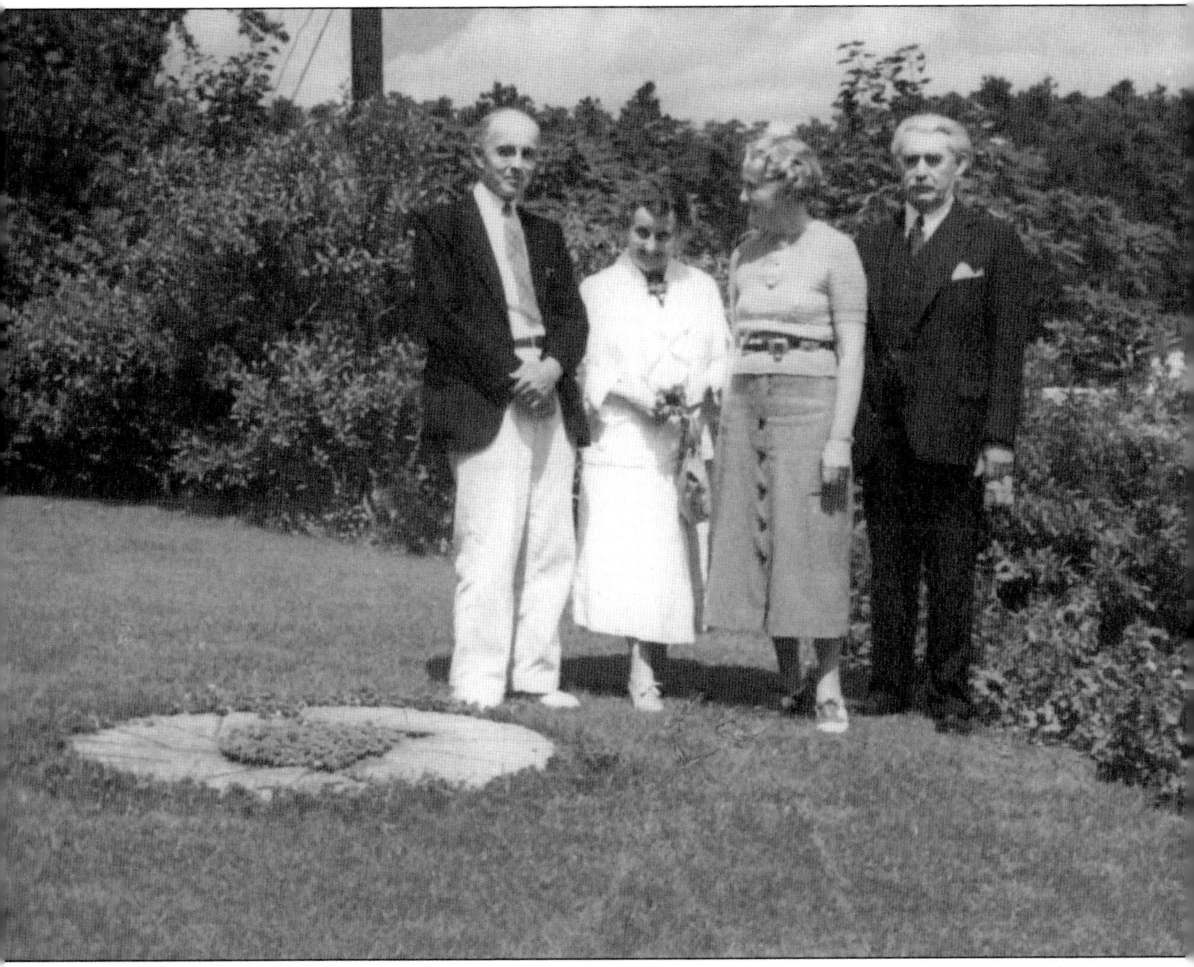

This family grouping is believed to include, from left to right, Cyril Farny; his mother, Sylvia; Cyril's wife, Phyllis Holt Farny; and Cyril's father, George. Cyril was 14 when the family moved to Craftsman Farms. In his 1984 book, *Cyril Farny: A 44 Year Pursuit of Color Photography*, he wrote about living at Craftsman Farms: "The house and the uniquely beautiful surroundings helped to further my emotional response to light and shadow, color, form, and texture." Scholar Marilyn Fish wrote about Cyril's teen years at Craftsman Farms, saying that he "seems to have single-handedly fulfilled Stickley's vision of practical, outdoor education for boys at the Farms by devoting many weekends and vacations to a variety of demanding agricultural chores." So it is very fitting that he and his wife, Phyllis, would be the ones to take over Craftsman Farms in the 1940s.

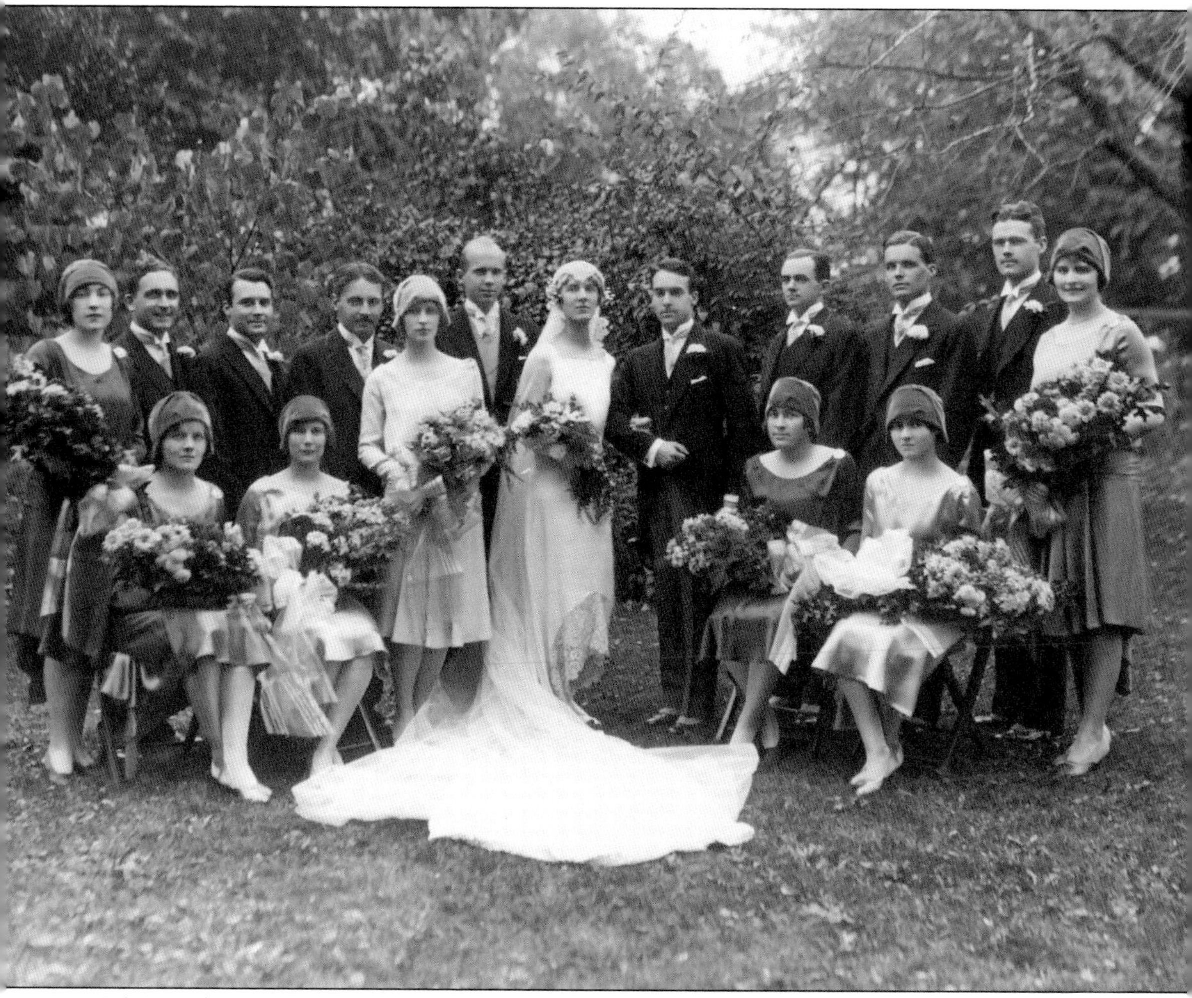

Sylvia and George's daughter, Alice Farny, was 12 when the family moved to Craftsman Farms. Alice began attending Kent Place School in Summit. There she met Phyllis Holt, and the two became best friends. They even attended Wellesley College together in later years. Phyllis visited Craftsman Farms frequently and was so close to the family that she called Major Farny "Uncle George." On these visits to Craftsman Farms, Phyllis met the Farnys' son Cyril, who was attending Princeton University. Phyllis and Cyril soon became a couple. This photograph of their wedding party was taken at the Holt estate in Summit, New Jersey, on October 7, 1928. Phyllis had just returned from her studies at the Frost School of Architecture in Oxford, England. Shown here from left to right are (first row) unidentified; Phyllis's two sisters, Evelyn and Ida Holt; Phyllis; Cyril; friend Harriet Hardy; and Eugene's wife, Margaret; (second row) Alice Wood; her husband, Richard Wood; unidentified; Eugene Farny; Rembrandt "Rem" Wurlitzer; unidentified; Phyllis's brother Robert Holt; unidentified; and Robert's wife, Kathleen.

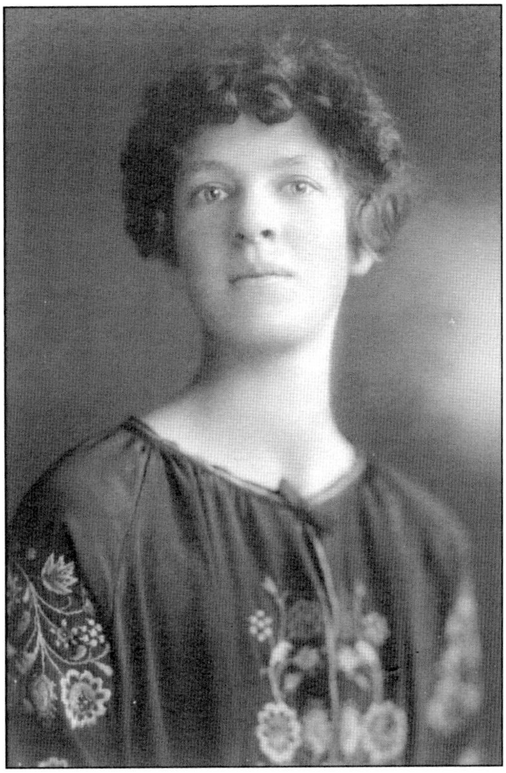

Phyllis Holt was born in Summit, New Jersey, in 1905, and seems to have been born to marry Cyril Farny and live at Craftsman Farms. She described her childhood home, which her father had built on the top of a hillside, as one with "wide lawns, grape arbors, fruit trees, vegetable and flower gardens, a tennis court, and a large shingled barn with chicken houses behind it. There were also pastures for the horses and cows, and plenty of woods with huge chestnut trees and tulip tress." Her father commuted to New York, like Gustav Stickley. Her mother loved music, like Sylvia and George Farny, and had even taught music at Elmira College in New York State. Phyllis is seen here in a portrait in the 1920s and later with her newborn son, Christopher. (Both, courtesy Ethel and Michael Farny.)

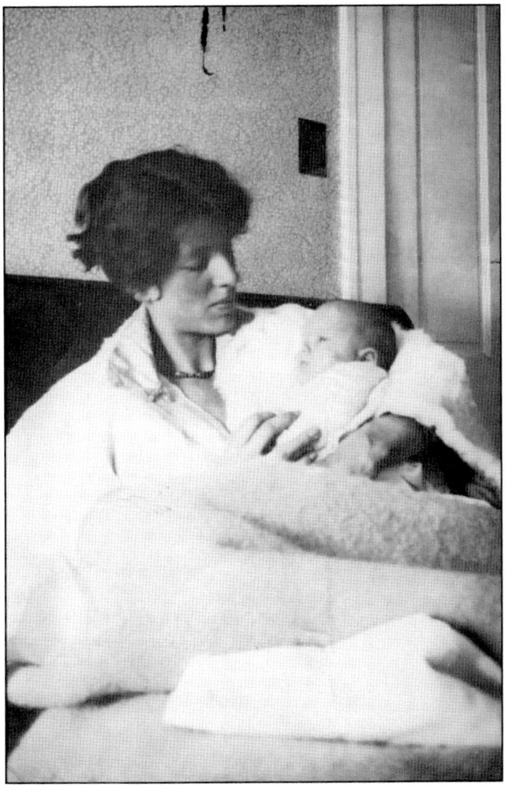

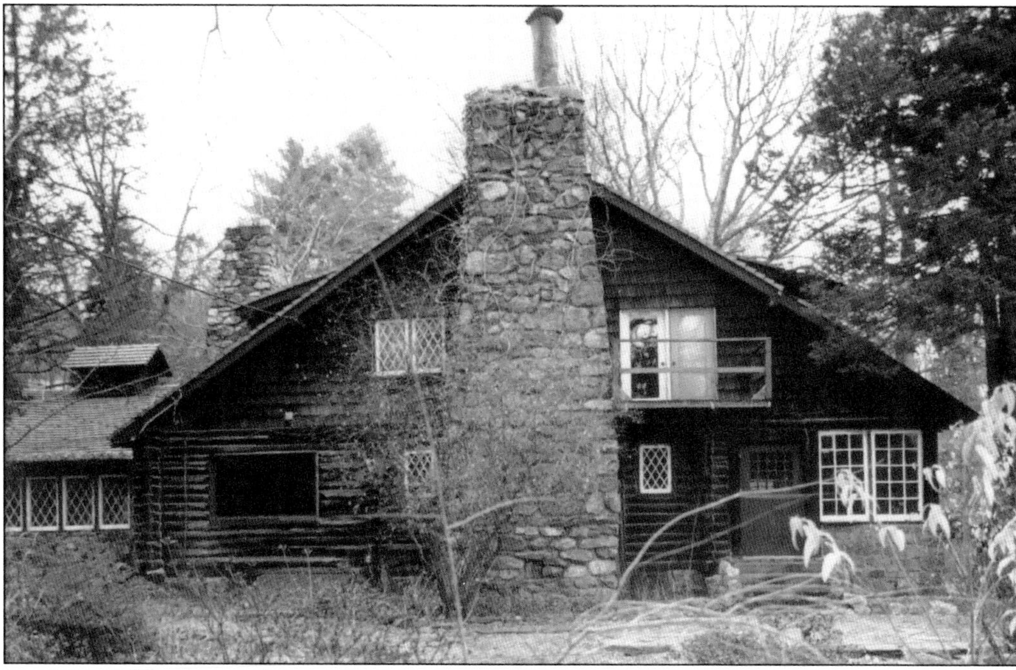

Cyril worked for the Wurlitzer Company's Piano Division, which was based in DeKalb, Illinois. He and Phyllis lived in DeKalb for more than 20 years while raising their four children: Christopher, Michael, Evelyn, and David. In the mid-1940s, after the death of Major Farny, Cyril and his family left Illinois to take over the family home, Craftsman Farms. Naturally, they made changes to the home to suit their needs and the esthetics of the time. Finding it quite dark inside, they installed a balcony off the south bedroom, which can be seen in these two photographs. It let in plenty of light while bringing a view of nature into the room.

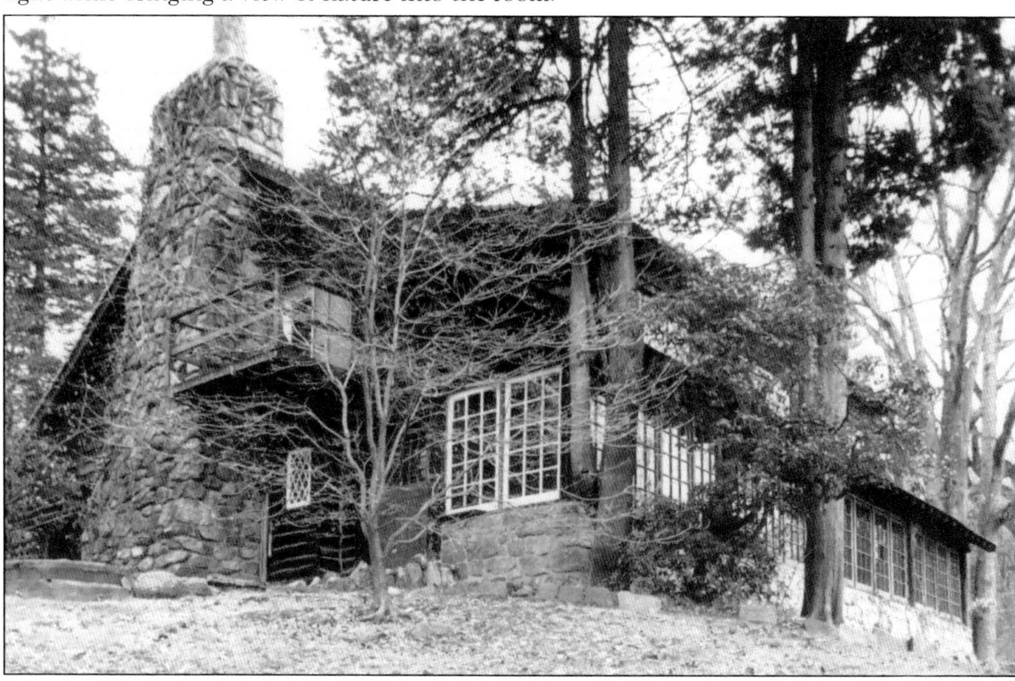

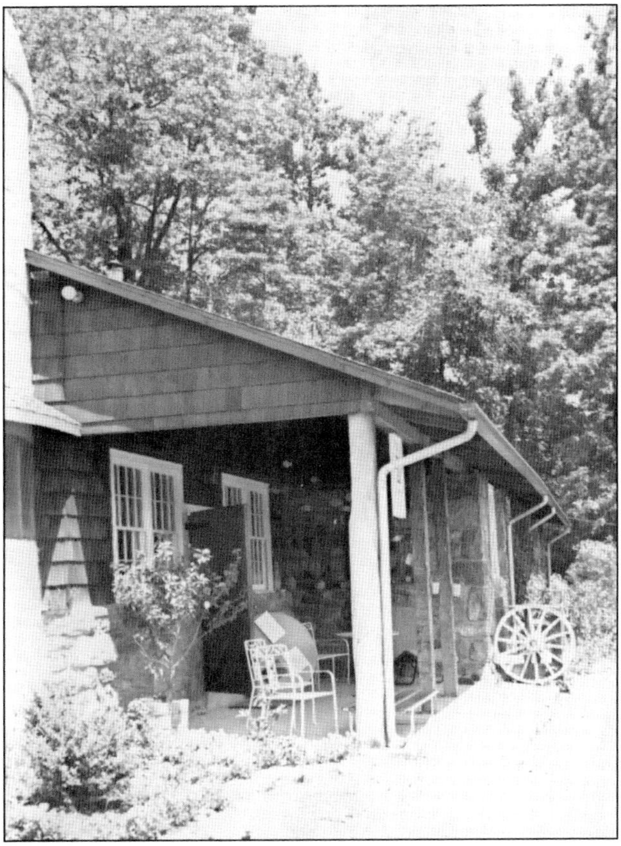

There are only a handful of known photographs of the cow barn, shown here with the silo that the Farnys added. For a while, Phyllis operated an art gallery and women's fashion store, which was named The Silo. Between 1955 and 1958, Cyril ran a commercial laundry equipment business, Laundramation, Inc. These images show the retail piano store they operated in the cow barn after the other business ventures. A terrible fire destroyed the building in the early 1970s. Today, the stone fireplace and some of its chimney remains are clearly visible but exposed to the elements.

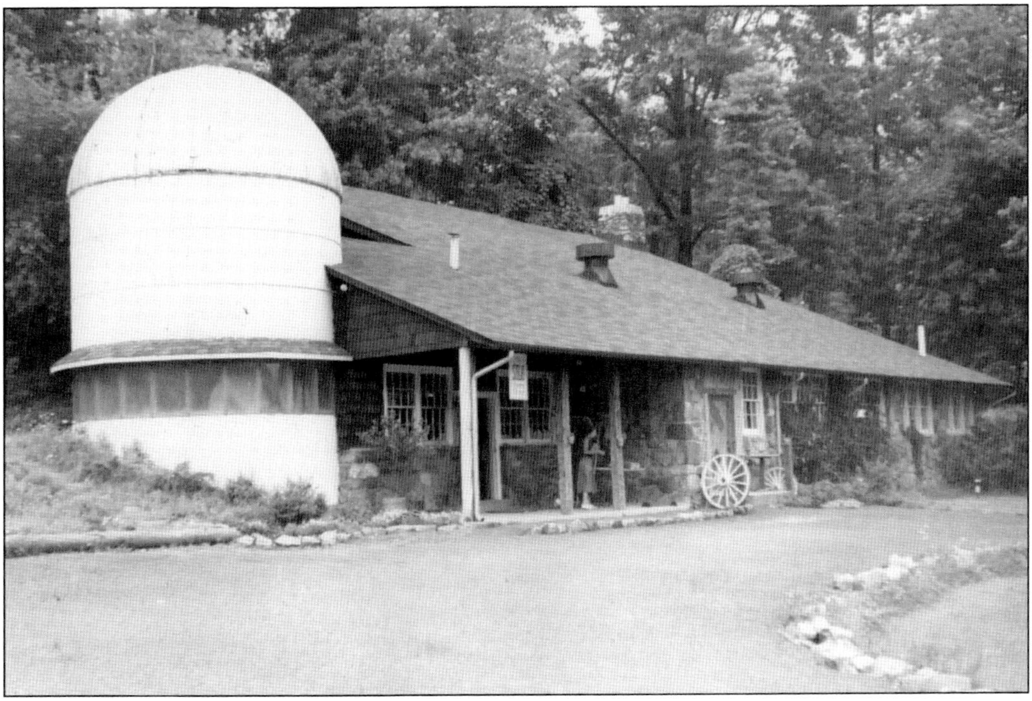

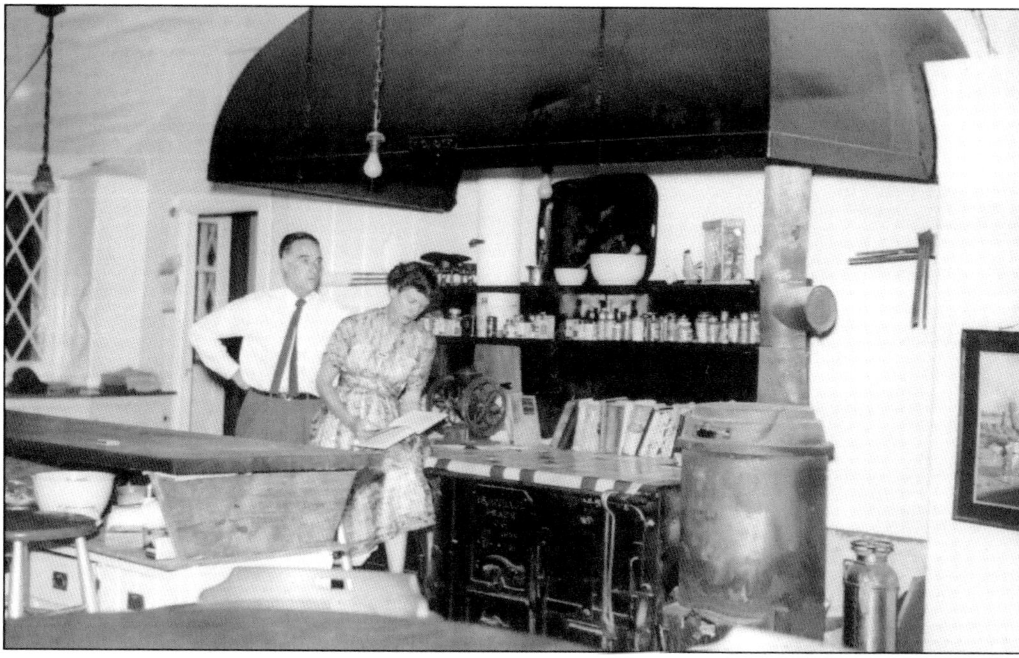

Cyril and Phyllis can be seen in the log house kitchen in the 1950s. The original Bramhall-Dean cast-iron range and massive hood can be seen in the back, thought its eight burners are covered with tiles for use as a counter top. What appears to be a woodstove for heating sits next to the range.

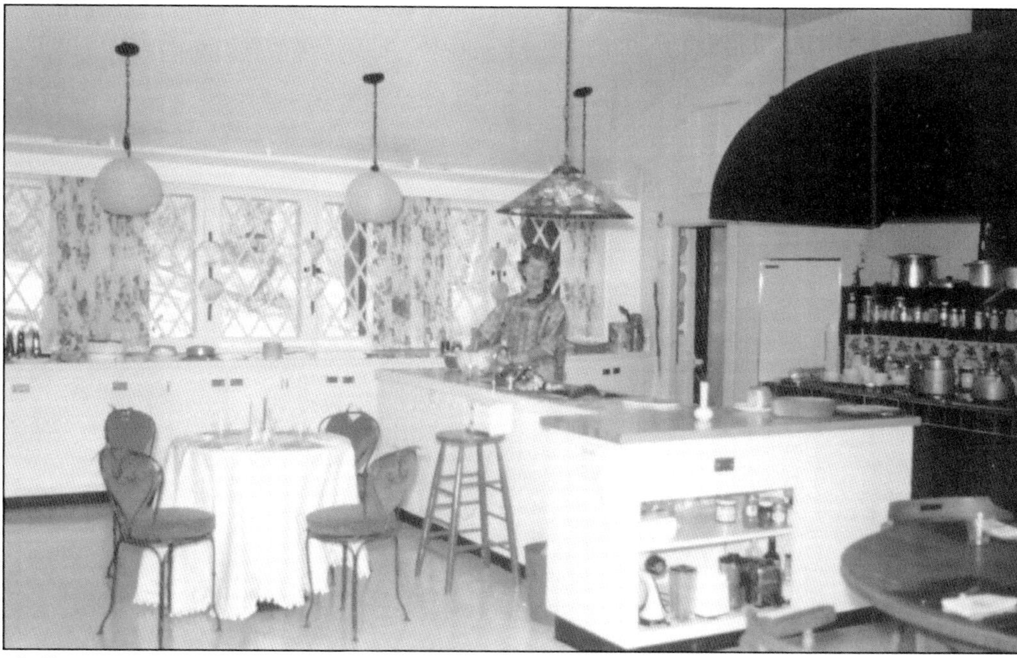

Phyllis is preparing a meal in this early-1960s photograph. The photograph shows the interesting mix of mid-century modern with original Stickley-era design. The original cast-iron range still remains. New modern-style light fixtures have replaced the bare bulbs seen in the photograph above. The original diamond-paned casement windows remain untouched.

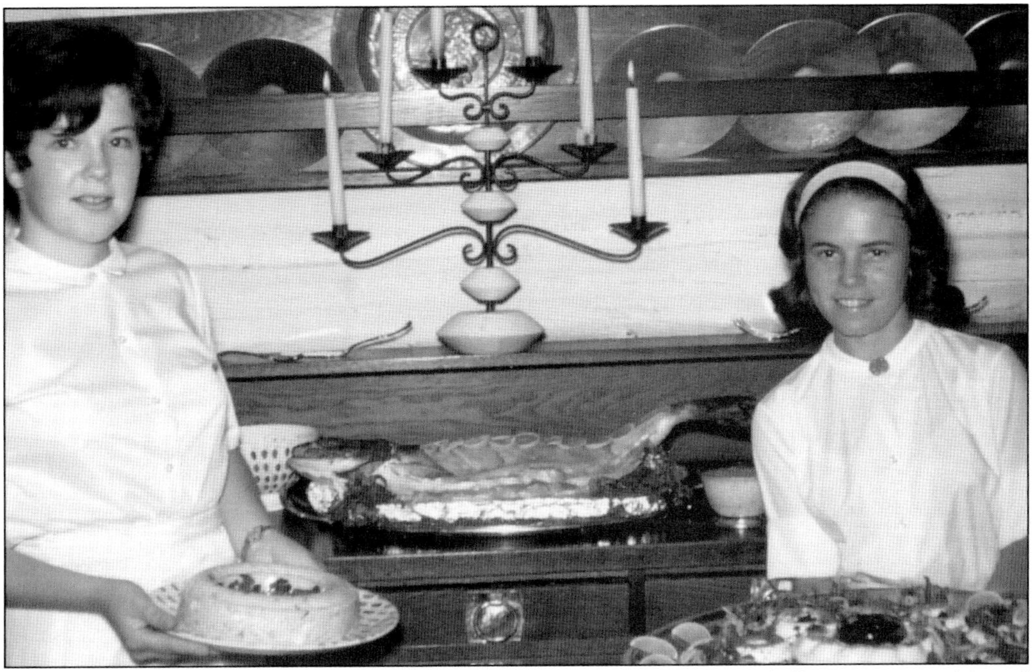

In the early 1900s, when electric lighting was new, most people distrusted illumination and feared that bright lights would harm their eyes. But by the time Cyril and Phyllis took over Craftsman Farms, its darkness was not seen as cozy but rather as depressing. Phyllis whitewashed the interior log walls and the fieldstone fireplace in an effort to make it feel fresh and bright inside. In the top photograph, two unidentified girls, believed to be servers at a party, pose with elegant party food. The log wall behind them is white, but the sideboard is still the same one that was in use when Gustav Stickley's family lived at Craftsman Farms. In the photograph below, white walls and white fieldstone are visible behind one of the carved wooden carousel horses collected by Phyllis.

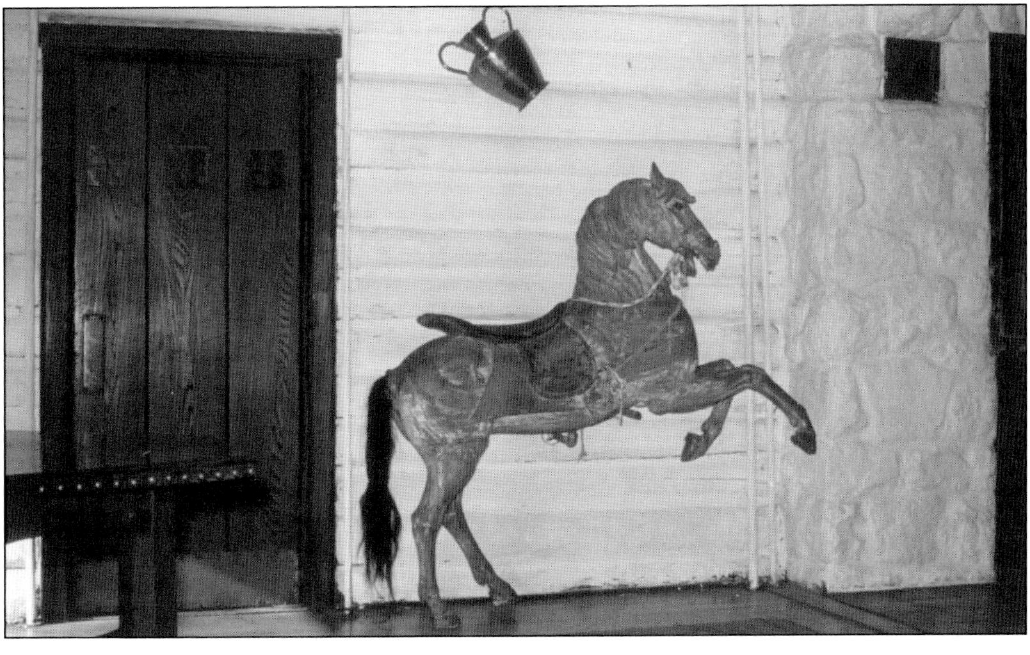

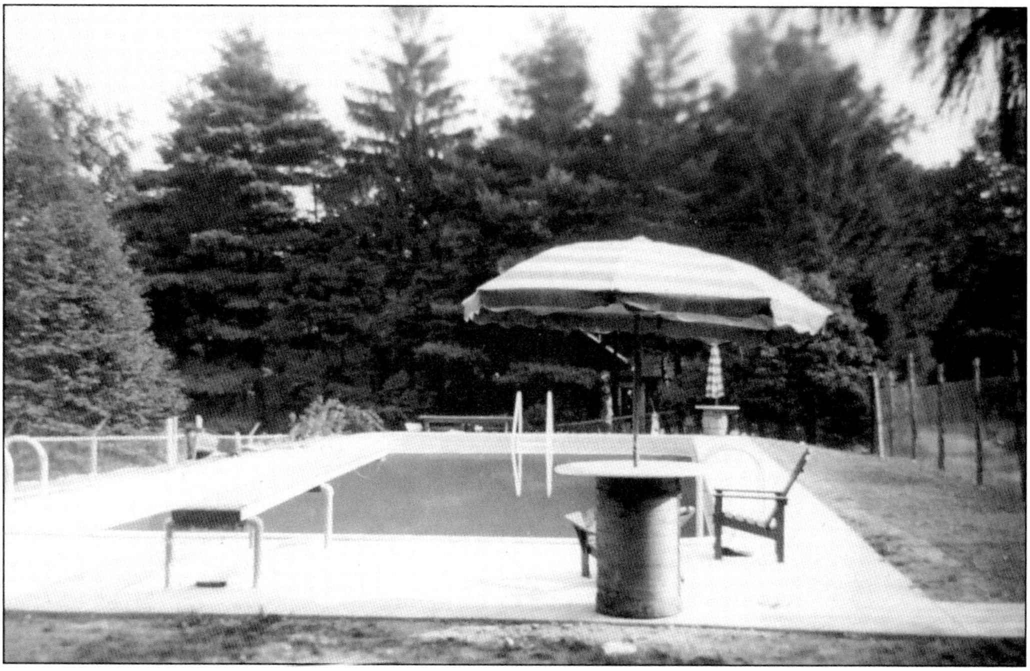

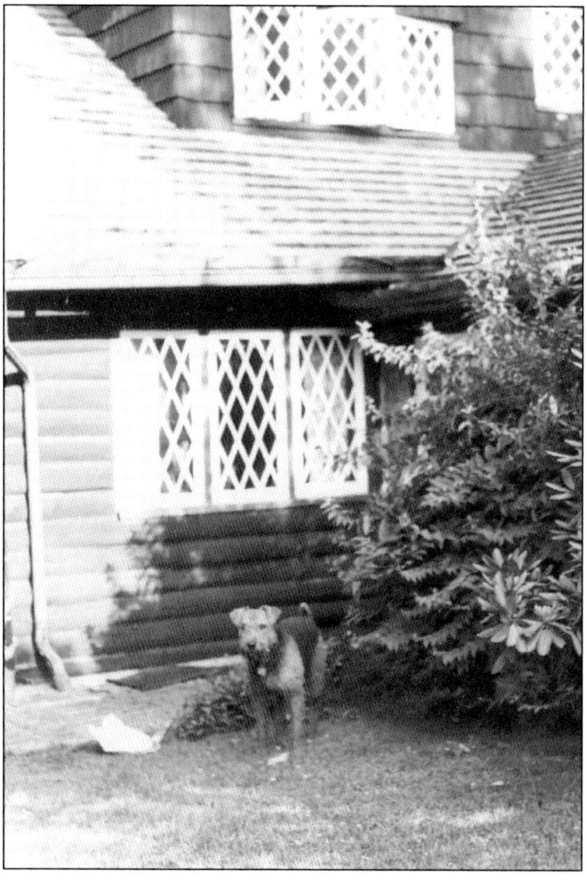

This 1960s photograph shows the metal pool the Farnys installed across the road from the servants annex. Over time, many of the farm buildings were converted into residential apartments. For a short period of time, these were operated as a private residential community called the Pigeon Hill Club. Members had access to the pool and the ski tow on the property.

Just as Sylvia and George loved Russian wolfhounds in earlier years, Phyllis and Cyril loved dogs. Their Airedale, Hector, is seen here playing on the lawn near the kitchen door. The foundation plantings, which were once small, have grown into such large shrubs that the kitchen door is completely hidden from view at this angle.

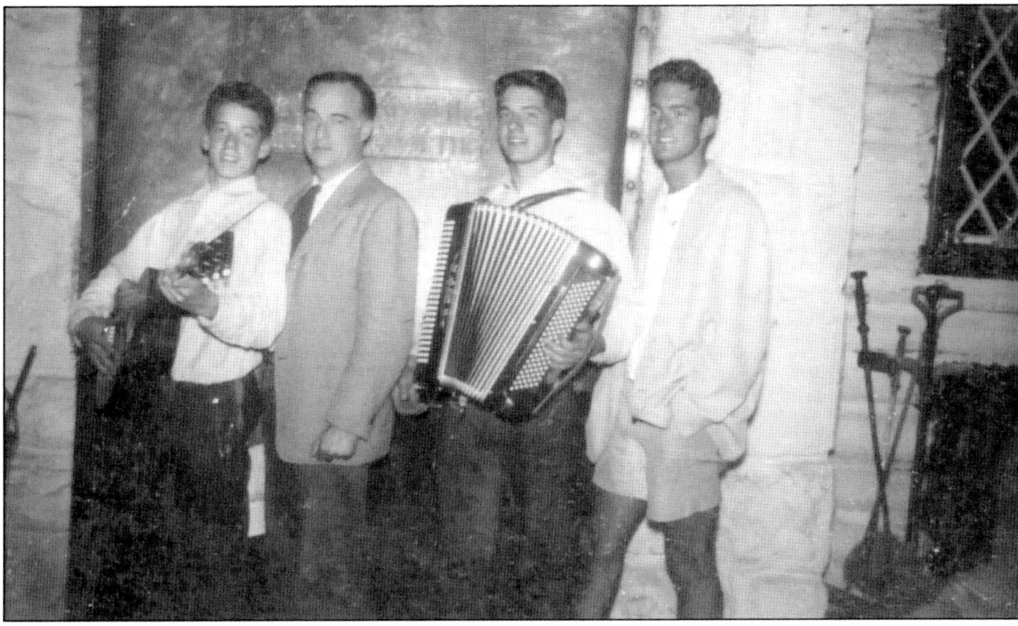

Music continued to be a big part of life at Craftsman Farms even on into the next generations. In the top photograph are, from left to right, Michael Farny playing the guitar; the boys' father, Cyril; David Farny playing the accordion; and Christopher. They are standing at the south end of the log house living room in the 1950s. The fireplace behind them has a hammered copper hood with the motto "the lyf so short, the craft so long to lerne." The photograph below was taken at a later date on the southern steps to the log house porch. Once again, Michael Farny is playing guitar and his brother David is playing the accordion. The other individuals are unidentified.

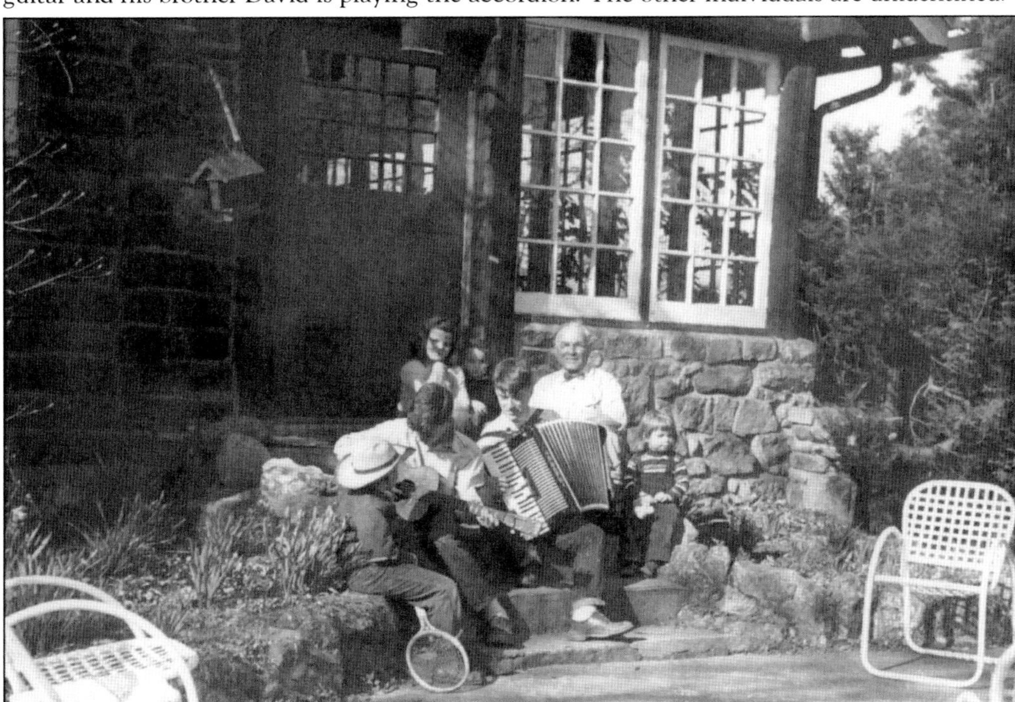

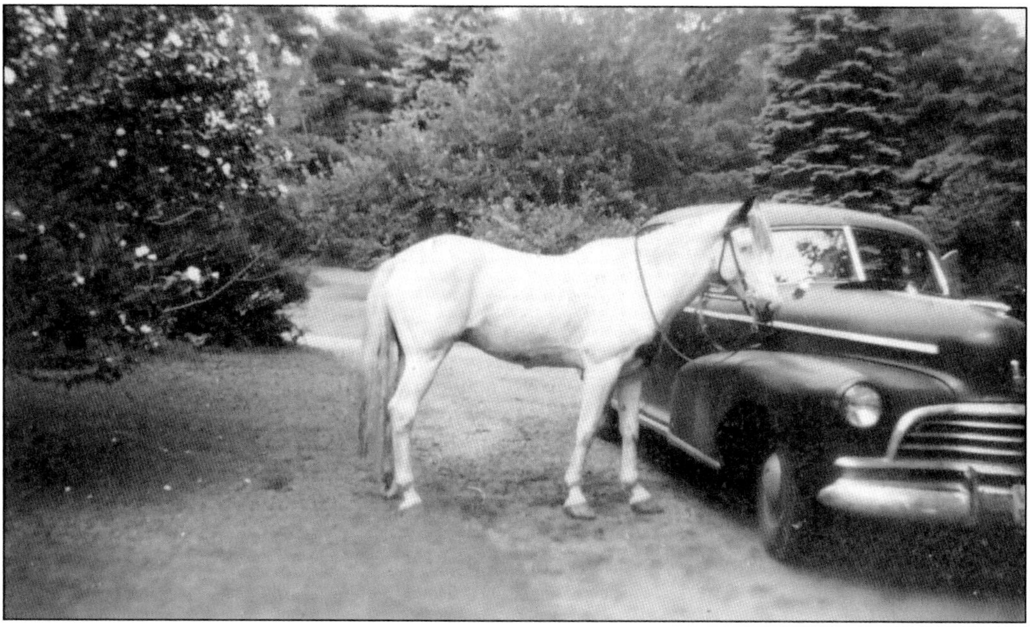

A white horse named Billy stands next to a family car on the north lawn of the log house in the early 1950s. Michael Farny lovingly described Billy as his "best friend." Family lore tells of the many tricks Billy would do, including crouching down so that Michael could climb on him.

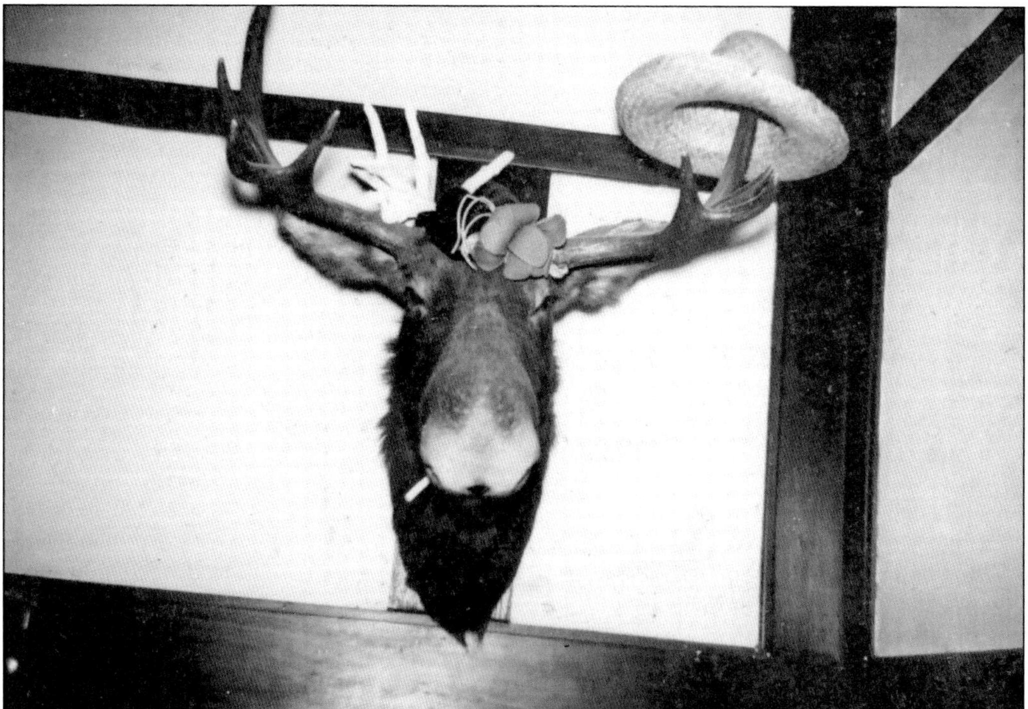

The rustic log cabin feel of the home at Craftsman Farms was mingled with a sense of humor during the Farny years. This moose head, which was displayed over the staircase in the log house, is decorated with felt flowers and a straw hat and holds a cigarette in its mouth.

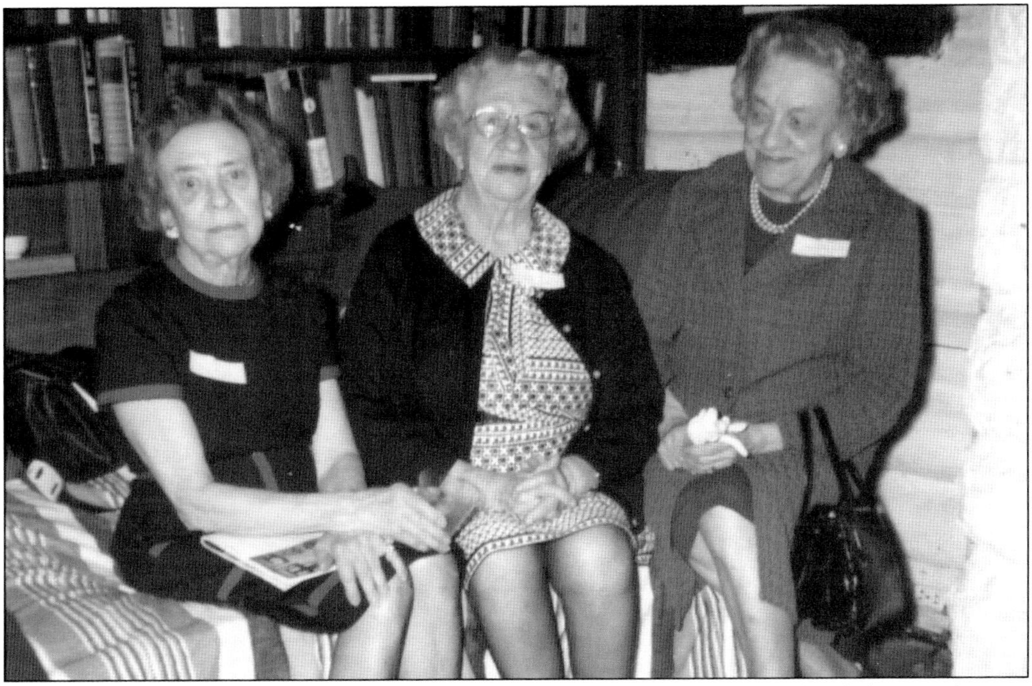

In 1972, Dr. Robert Judson Clark curated Princeton University's landmark exhibition *The Arts and Crafts Movement in America, 1876–1916*. He was fond of the Farnys, whom he had known for about a decade. In 1973, when Cyril and Phyllis decided to retire to Colorado, Clark helped them plan a farewell party. It was a reunion of several generations of people who had lived at Craftsman Farms or were associated with it since its beginnings. The top photograph is of, from left to right, Hazel Stickley, Barbara Stickley Wiles, and Mildred Stickley Cruess at the party on May 5, 1973. The bottom photograph of the log house at the end of the Farnys' time shows the plantings, which had grown vigorously and were still maintained. Inside the log house, most of the original Stickley furnishings remained in place.

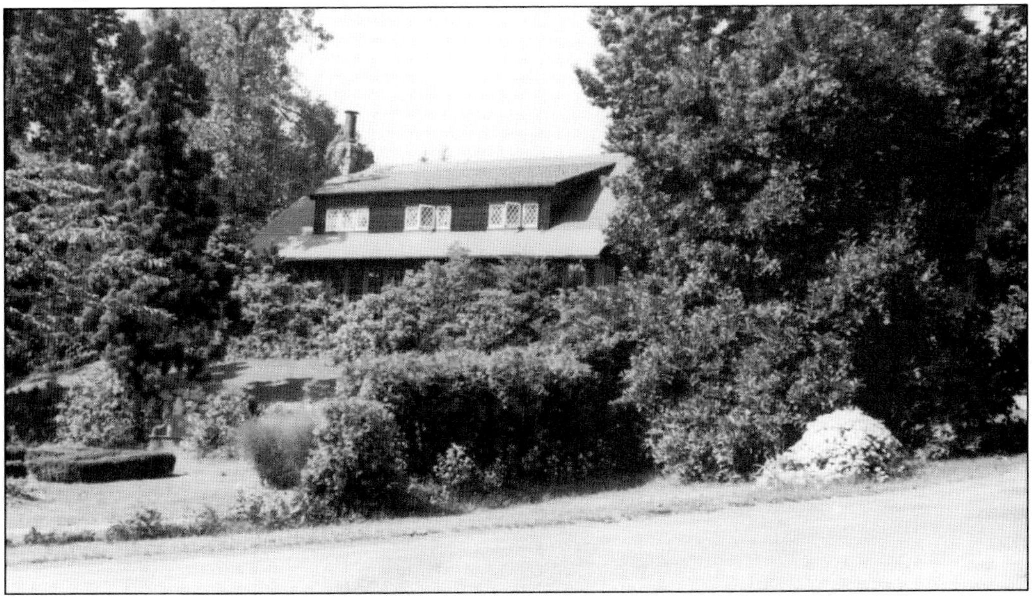

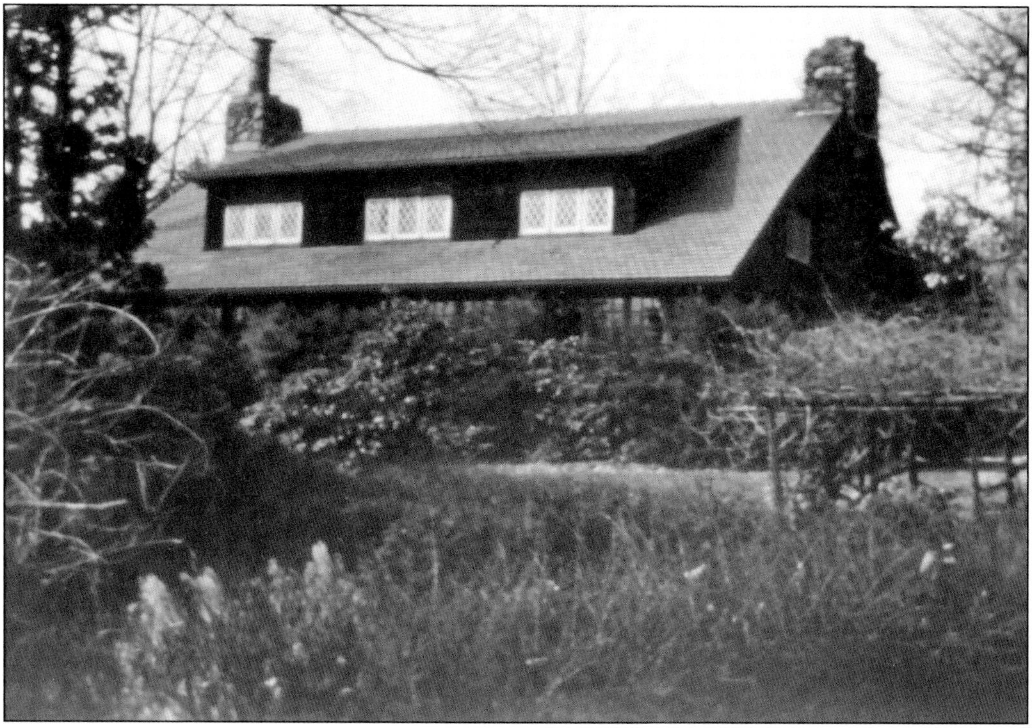

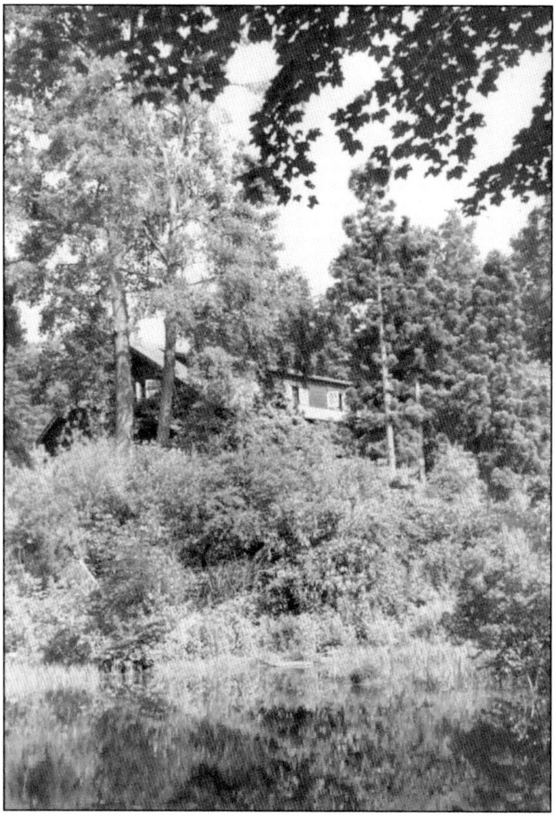

Phyllis and Cyril's children did not wish to take over the estate, so J. Harry Du Bois, a retired engineer that Cyril had known from the Wurlitzer Company, was hired as the resident agent. Harry and his wife, Kathryn, lived in the log house and cared for the estate until 1984, when Donald Straits became the agent. Phyllis died in 1975. Using advertisements in historical publications, Cyril continued his quest to sell Craftsman Farms to a buyer who would keep the remaining parcel intact. These undated photographs of the log house show how the landscape changed during this time. Dr. Robert Judson Clark wrote, "Not long afterward the Farnys moved to Colorado . . . the fate of the place became frighteningly precarious. Then came the Craftsman Farms Foundation in 1989, and the great rescue began."

# *Six*

# Rescue and Rebirth

For more than a decade after Cyril and Phyllis Farny retired to Colorado, Craftsman Farms continued to be managed as small rental units. Over the years, more than 600 acres of Stickley's pastureland, farmland, and woodland were sold and developed. Miraculously, the central 26 acres containing the log house and cottages still survived. The Stickley-era farm structures, except the barn, which was destroyed by fire, also survived in another small tract of land sold to Phyllis's brother in December 1985.

With property taxes of approximately $35,000 a year and mounting maintenance costs, the need to find a buyer was urgent. Eventually, Cyril Farny signed a contract to sell his 26 acres to Foreston Development Corporation of New York. The contract stipulated that there could be no substantial alterations to the log house and was contingent upon the developer's ability to obtain a variance to build 52 townhouses. Architect Robert A.M. Stern was hired to design these townhouses in a Stickley style.

A group of preservationists and Arts and Crafts aficionados mounted a grassroots campaign to save Craftsman Farms. These vocal advocates garnered a great deal of publicity about the historical importance of the site. The remaining parcel of land was located in the Township of Parsippany–Troy Hills, and the township officials took note of the mounting concern for the property.

In March 1987, the Township of Parsippany–Troy Hills applied for $1.5 million in Green Acres Loan funds to buy Craftsman Farms. In April 1988, it collaborated with Morris County Trust for Historic Preservation, the New Jersey Historic Trust, and the National Trust for Historic Preservation to commission a feasibility study to determine appropriate uses of Craftsman Farms. The study recommended that the township acquire Craftsman Farms with the further recommendation that the site be managed by a nonprofit corporation. In November 1988, Parsippany voters overwhelmingly approved the establishment of an Open Space Trust Fund to purchase the property. Finally, in 1989, the township used eminent domain to purchase the property and the nonprofit Craftsman Farms Foundation was formed to manage Craftsman Farms as a historic site.

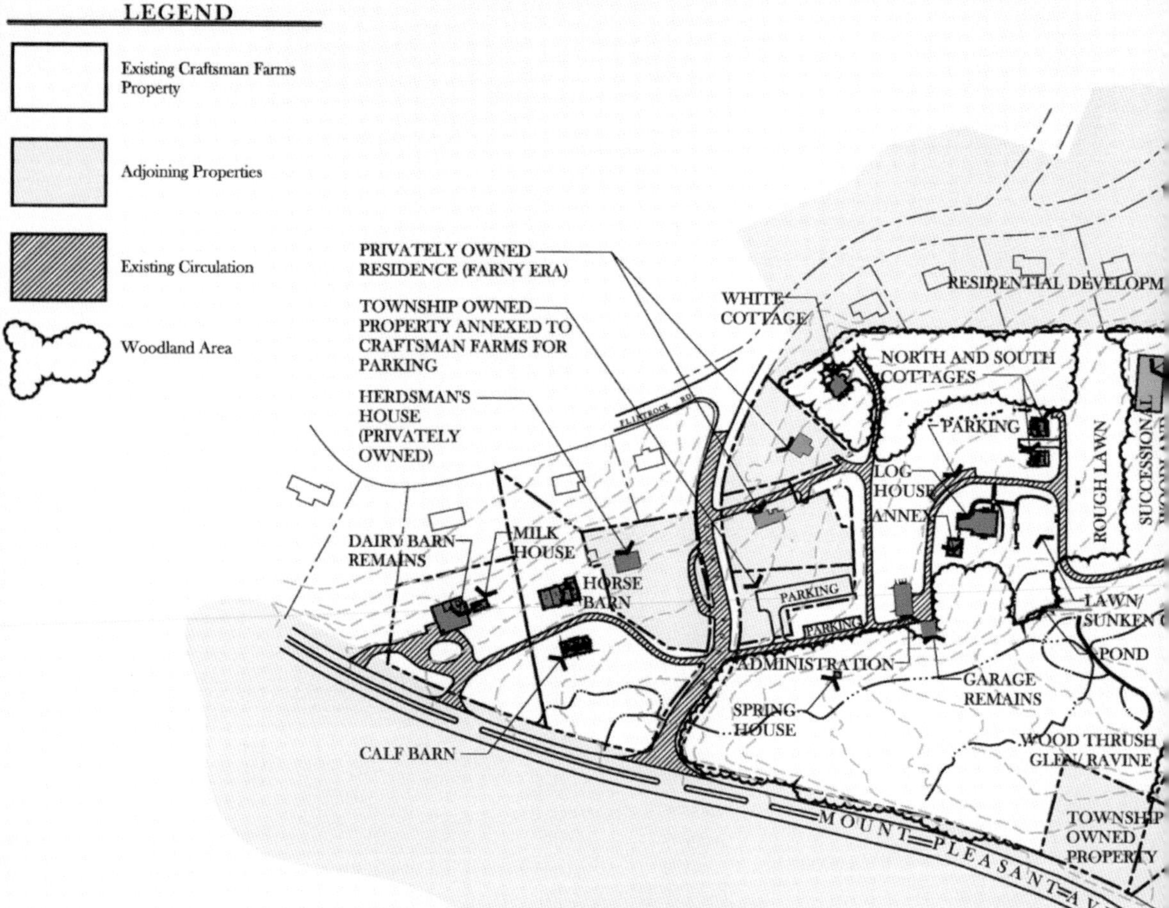


## Map Sources:

Craftsman Farms (wetlands map)
(Lot 31, Block 25, Tax Map Sheet 34)
Township of Parsippany - Troy Hills, Morris County, New Jersey

"Craftsman Farms" Temple University
Stickley Period c. 1916
Morris Plains, New Jersey
Landscape Architect and Horticulture, Fall Junior Studio 1994
Sara J. Hedstrom

Archeology.jpg -Scan
Hunter Research
120 West State Street
Trenton, NJ 08608
609-605-0122

Google Earth Aerial
Image © 2007 State of New Jersey
© 2007 Europa Technologies
© 2007 Tele Atlas

Craftsman Farms Existing Site Plan
March 6, 1992
School of Architecture, New Jersey Institute of Technology
Advanced Options Studio, 1992

Craftsman Farms - First Floor Plan (Log House)
May 9, 1992
School of Architecture, New Jersey Institute of Technology
Advanced Options Studio, 1992

"Roots In the Ground" A historic Landscape Study of Craftsman Farms
Parsippany - Troy Hills, New Jersey
Site In 1998, Drawing 2
Mary K. Muckenhoupt
April 1998

"Roots In the Ground" A historic Landscape Study of Craftsman Farms
Parsippany - Troy Hills, New Jersey
Core Area Vegetation, Drawing 4
Mary K. Muckenhoupt
April 1998

Craftsman Farms Tree Identification
Done Winter of 1992
By: J S L Greentree, Landscape Architect
Sheep Hill Rd.
Boonton Twp., NJ 07005
Field Assistance: Lorraine Caruso, Botanist

Topographic Map, Township of Parsippany - Troy Hills
Morris County, New Jersey
Compiled By: International Resources and Geotechnics, Inc.
Contour Internal = 2', December 1972
Sheet No. 3652

Topographic Map, Township of Parsippany - Troy Hills
Morris County, New Jersey
Compiled By: International Resources and Geotechnics, Inc.
Contour Internal = 2', March 1975
Sheet No. 3649
Tax Map, Township of Parsippany - Troy Hills - Sheet 34
Morris County, New Jersey, October 2001

Glen J. Lloyd, L.S., Schoor Depalma
160 Littleton Rd. Suite 300
P O Box 5245
Parsippany, NJ 07054

Tax Map - Township of Parsippany - Troy Hills, Sheet 35
Morris County, New Jersey
December 1971
International Resources and Geotechnics, Inc.
Westchester County Airport
White Plains, NY

Architectural Building Floor Plans for:
White Cottage, Chicken Coop, Dairy Barn, Horse Barn, Garage, North Cottage, South Cottage, Offices, Annex
Mark Hewitt Architects
114 Claremont Rd.
Bernardsville, NJ 07924

Craftsman Farms 1938 Aerial - No Source Information

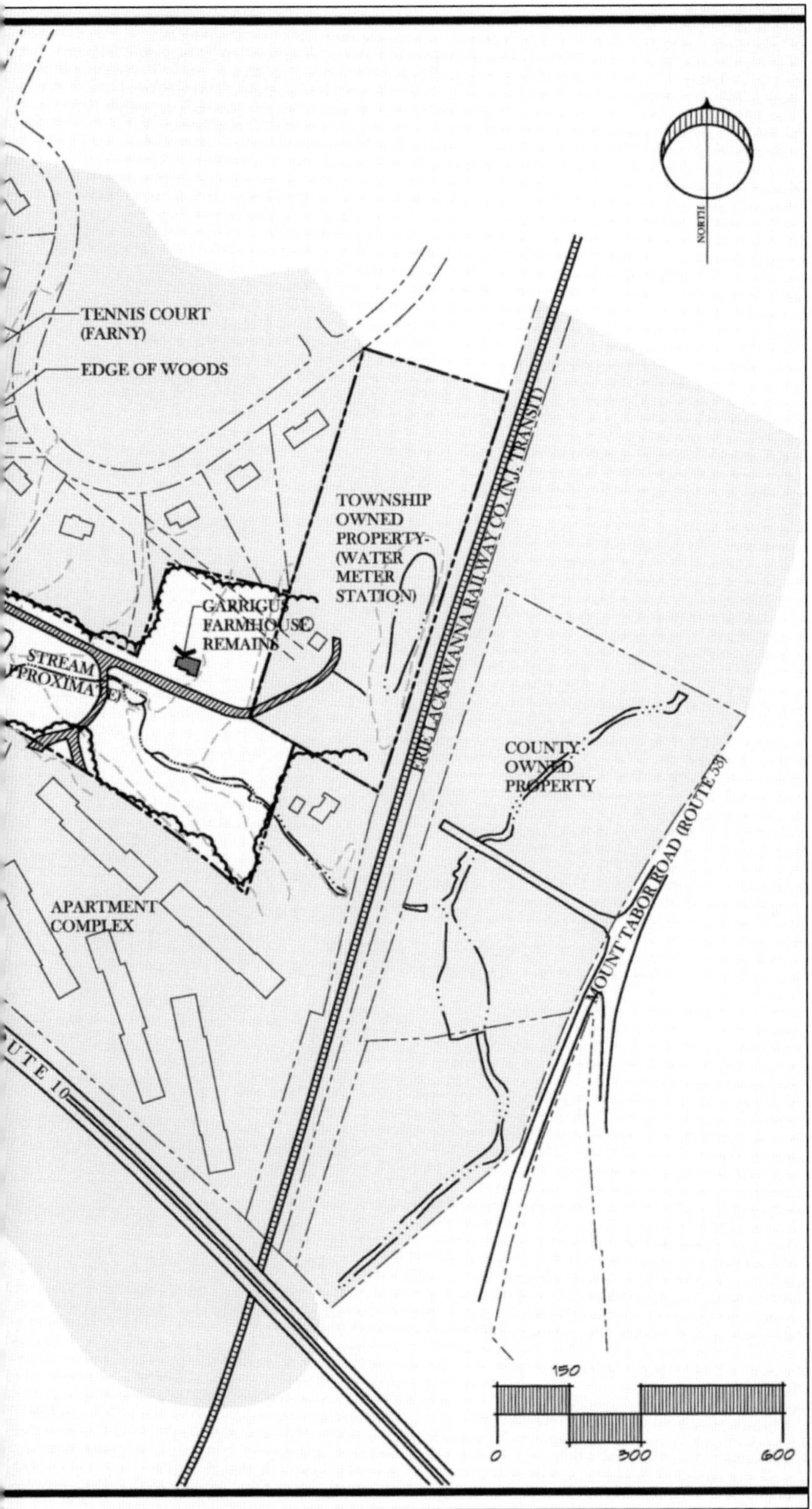


This map of the Craftsman Farms today shows that though only 30 acres remain of the original 650 acres, the most important part of the property remains intact. All the original Stickley-era buildings remain intact except the cow stable, which burned in the 1970s, and the upper floor of the workshop garage, which burned in 1950. The original buildings experienced some architectural alterations but are gradually being returned to their original appearance. The few structures added by the Farnys were sensitively built and relate well to the Craftsman style. The overall landscape form has not been altered enough to prevent its future restoration. (Courtesy B.W. Bosenberg & Co., Inc.)

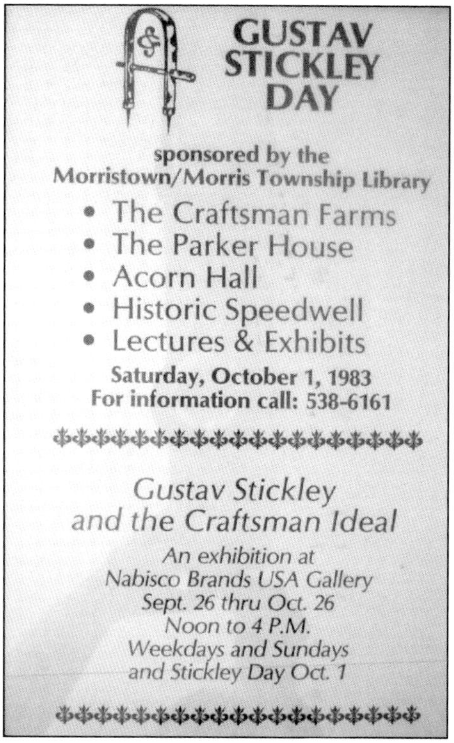


The first major event to draw attention to the historical significance of Craftsman Farms took place on October 1, 1983: Gustav Stickley Day. It was sponsored by the Joint Free Public Library of Morristown and Morris Township. Four historic places were opened to the public, including Craftsman Farms and the nearby Parker House, which was also designed by Gustav Stickley. (Courtesy Barbara Fuldner.)

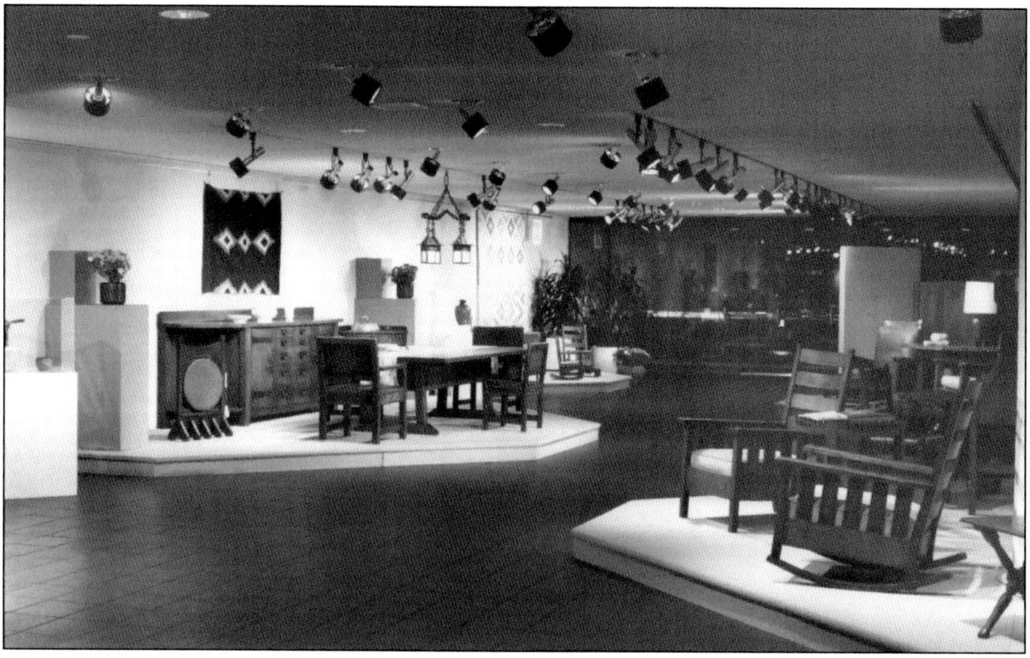

In conjunction with Gustav Stickley Day, a month-long exhibition entitled *A Celebration of Quality: Gustav Stickley and the Craftsman Ideal* was presented at the Nabisco Brands USA Gallery in East Hanover, New Jersey. Produced by the New York City arts promotion firm of Naramore and Stein, the exhibition, which is pictured here, drew national attention and attracted visitors from 21 states.

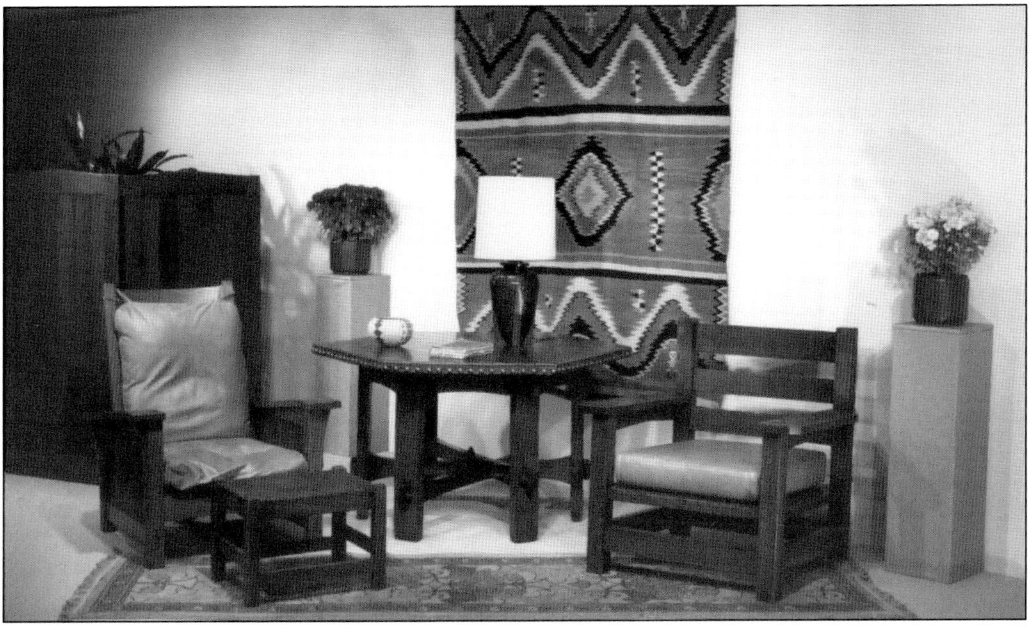

Elaine Hirschl Ellis (then Elaine Naramore) and Anthony Stein of Naramore and Stein borrowed Stickley objects from collectors, museums, and the Jordan Volpe Gallery. Ellis arranged the items in room settings, as Stickley had done in exhibiting his furnishings years before. The exhibition was a pivotal event, raising the profile of Craftsman Farms and making a strong case for historic preservation.

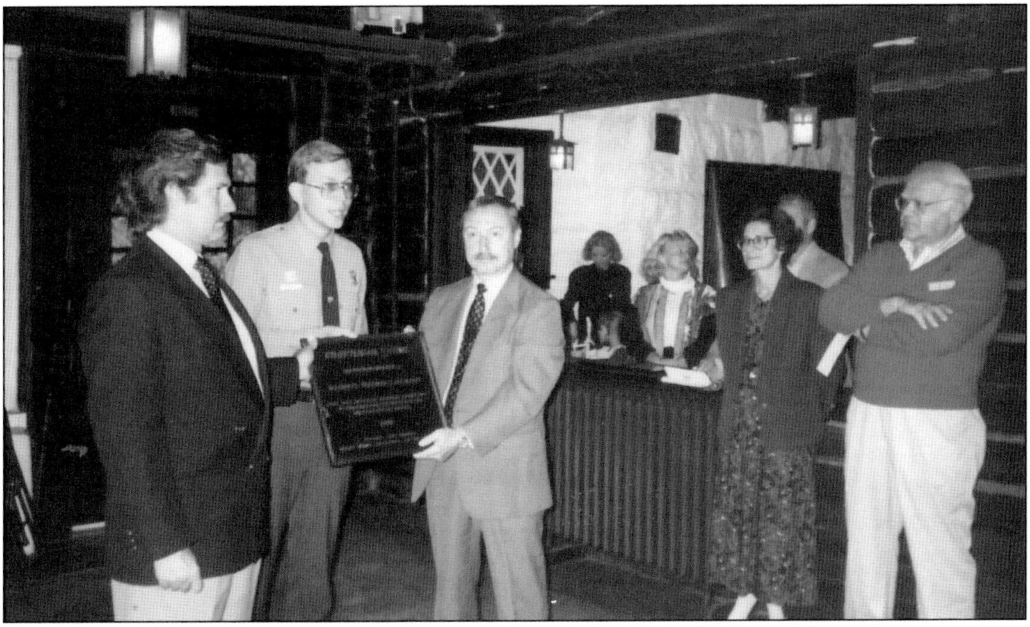

In 1989, Craftsman Farms was named to the National Register of Historic Places and nominated for National Historic Landmark status. Shown at the celebration of the site's 1990 designation as a National Historic Landmark are, from left to right, Mayor Frank Priore; unidentified; Councilman Michael dePierro; Craftsman Farms Foundation's director, Vivian Zoe; and founding chair of the Morris County Open Space Committee Dr. Joseph Weisberg.

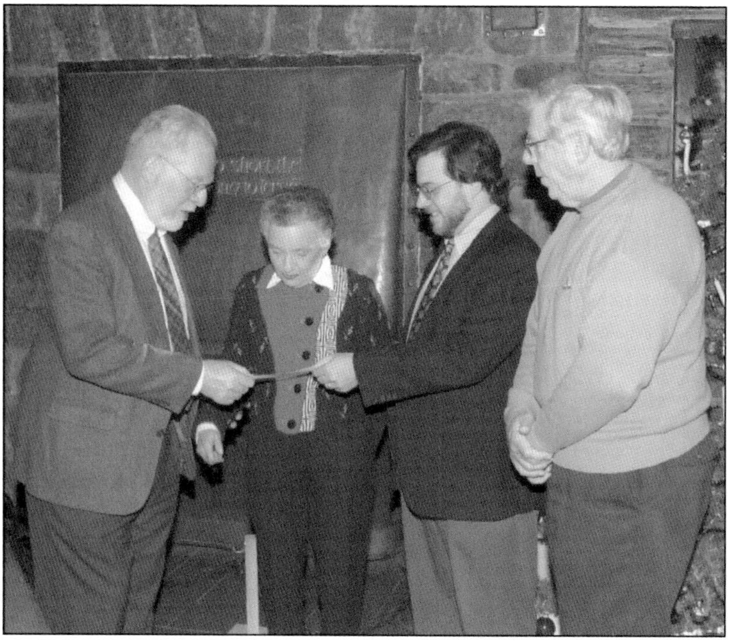

On of the first businesses to support Craftsman Farms was Bograd's Fine Furnishings. In 1990, Bograd's featured newly reissued Gustav Stickley designs produced by the Stickley-Audi Company and hosted a benefit for Craftsman Farms. This 2004 photograph shows one of its many donations. From left to right are Joseph Bograd; Marcia Bograd; Craftsman Farms Foundation's executive director, Tommy McPherson; and trustee Davey Willans.

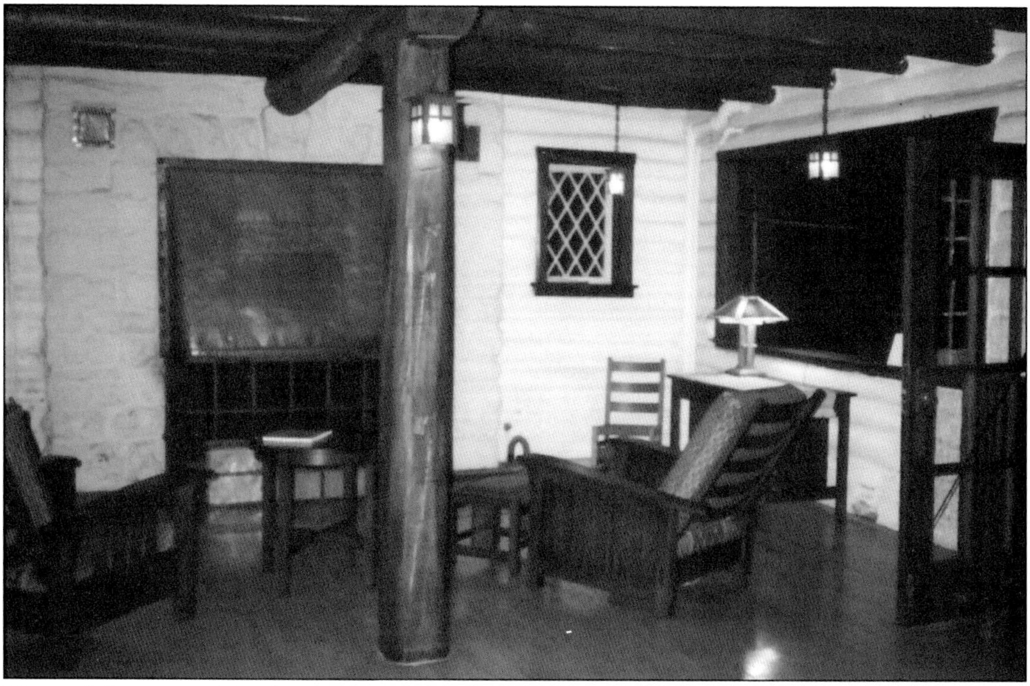

Most of the original log house furnishings had been sold at a Christie's auction. In this 1990 photograph, the living room was furnished with loans of antiques combined with reissued Stickley furnishings. That year, the house was opened to the public twice a week from July through October.

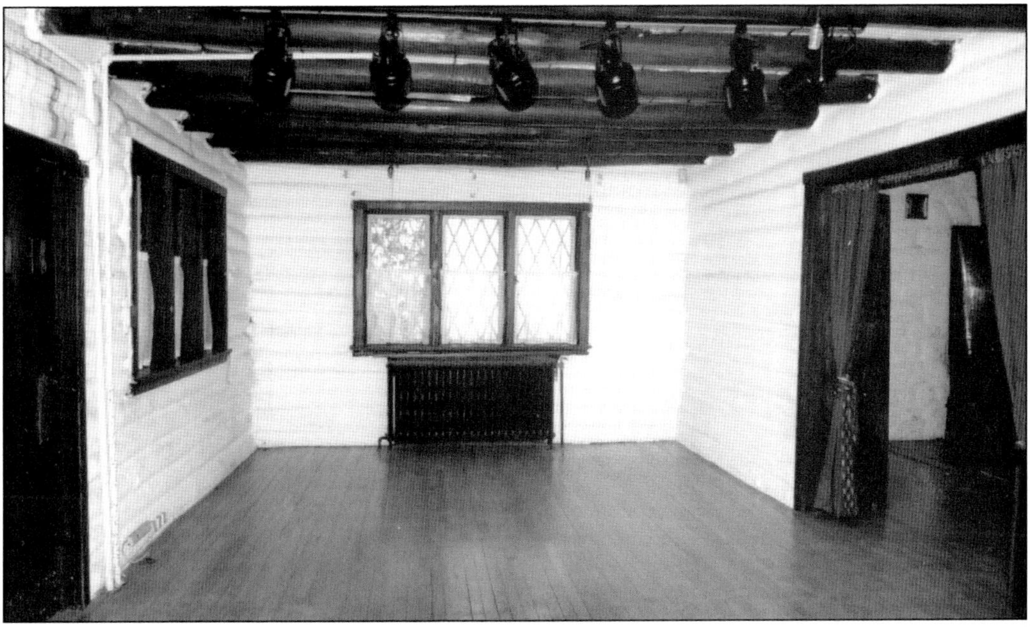

The New Jersey Historic Trust awarded the Craftsman Farms Foundation a matching grant of $100,000 to undertake a historic structures report on the log house and to begin the first phase of restoration. The firm of Holt Morgan Russell (HMR) Architects was selected to do the work. It was later discovered that Philetus H. Holt III, principal at HMR, was the nephew of Phyllis Holt Farny. He had even stayed overnight at the log house as a young man. These two images of the log house in 1993 were taken by HMR as part of the log house conditions study. The top one shows the north end of the dining room with the painted walls. The white painted fireplace stones can be seen through the doorway on the right. The bottom photograph shows the staircase with Stickley's monogram "S" cut out of the railings. With the completion of the historic structures report, the Craftsman Farms Foundation took the first steps on the long road to restoring the log house to its 1911 appearance.

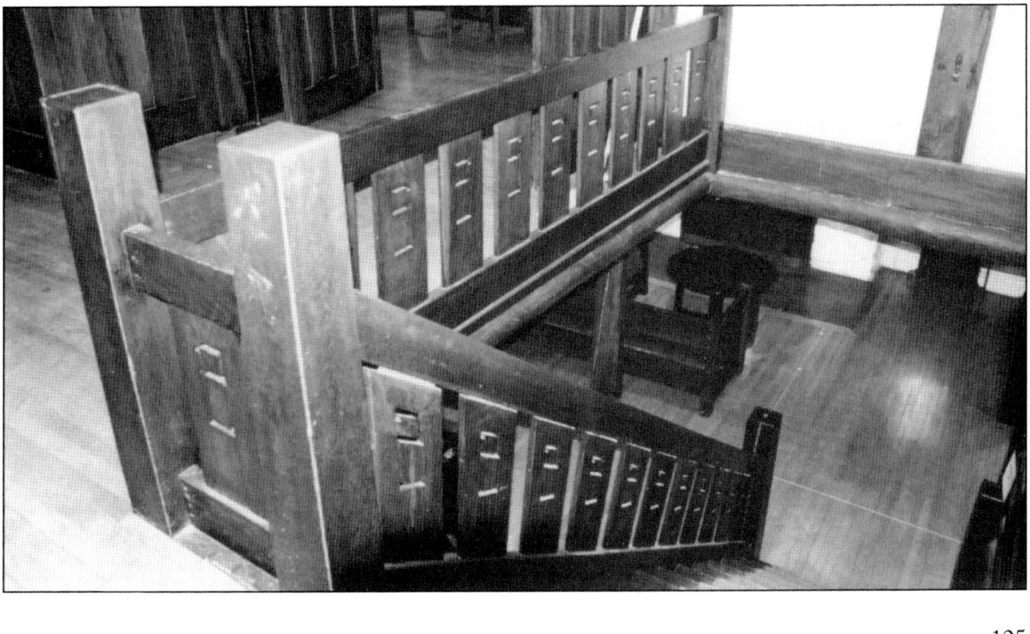

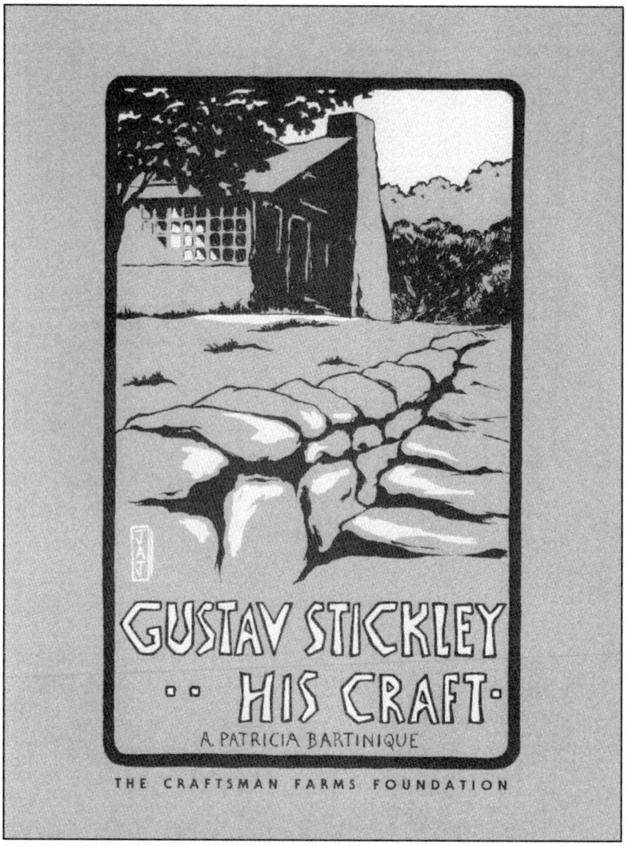


The Craftsman Farms Foundation presented its first major catalogued exhibition, *Gustav Stickley: His Craft*, from November 15, 1992, to January 31, 1993. It was curated by A. Patricia Bartinique and was described as a comprehensive exhibition of Stickley's work. Bartinique wrote a large portion of the accompanying catalogue, shown here. The catalogue, which also included essays by 14 other Arts and Crafts authorities, has become a resource for Arts and Crafts enthusiasts and scholars. The image below shows a party celebrating the success of the exhibition that was held for the Craftsman Farms volunteers. From left to right are Nancy Willans; Muriel Berson, who was the most vocal advocate for saving Craftsman Farms; Nancy Strathearn, who was named executive director that year; A. Patricia Bartinique, who curated the exhibition; Judith Jordon; and Jeanne P. Jones.

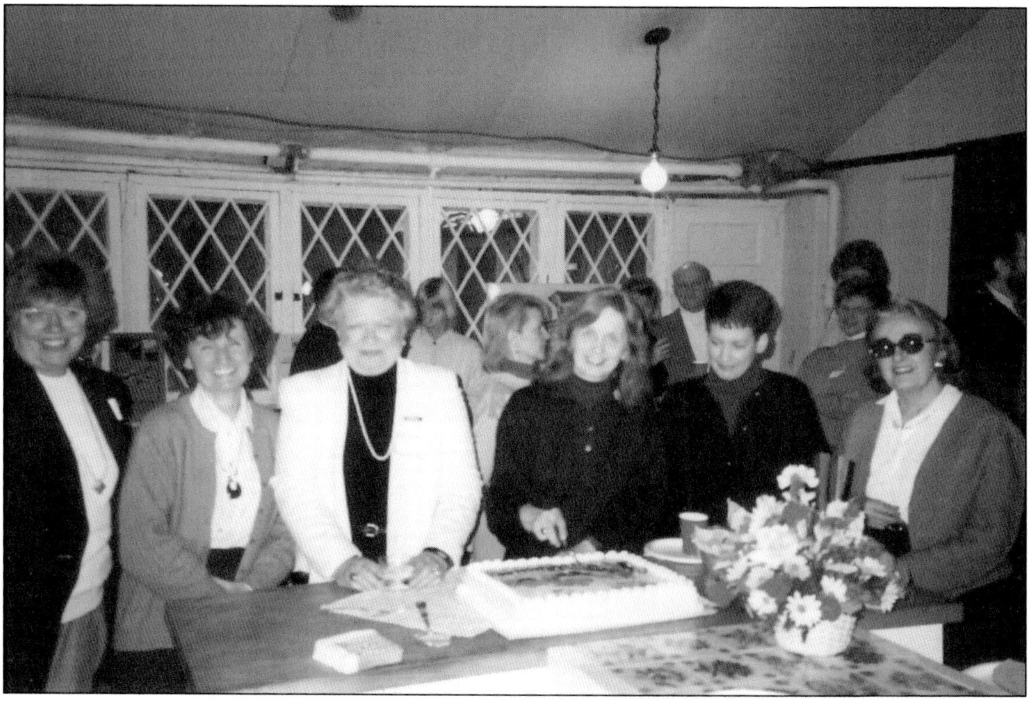

The historic structures report had urged that the log house roof be rebuilt as soon as possible. The Craftsman Farms Foundation used major preservation grants from the New Jersey Historic Trust and the Morris County Preservation Trust along with individual donations to pay for the roof. This flyer was used to drive individual donations.

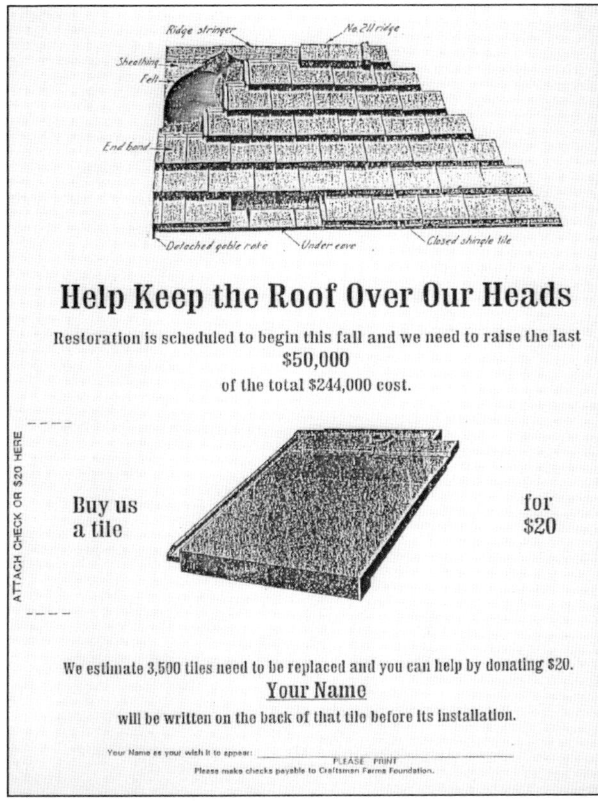


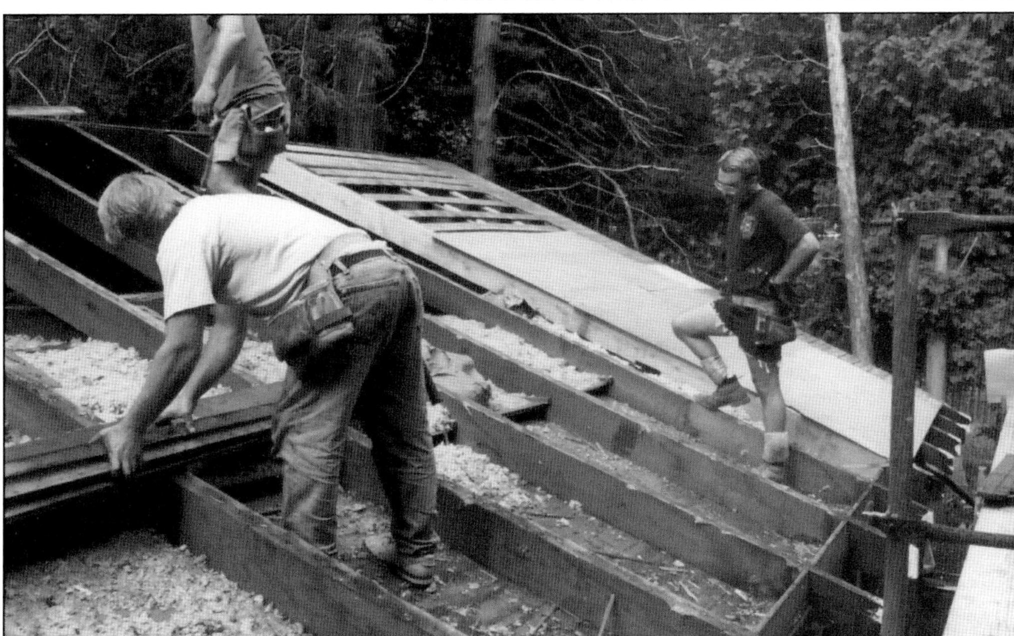

Workers have stripped the roof over the kitchen wing down to the bare rafters in this 1994 photograph. Following the Secretary of the Interior's Guidelines for Historic Preservation was arduous but was a wonderful step towards preserving the log house for future generations.

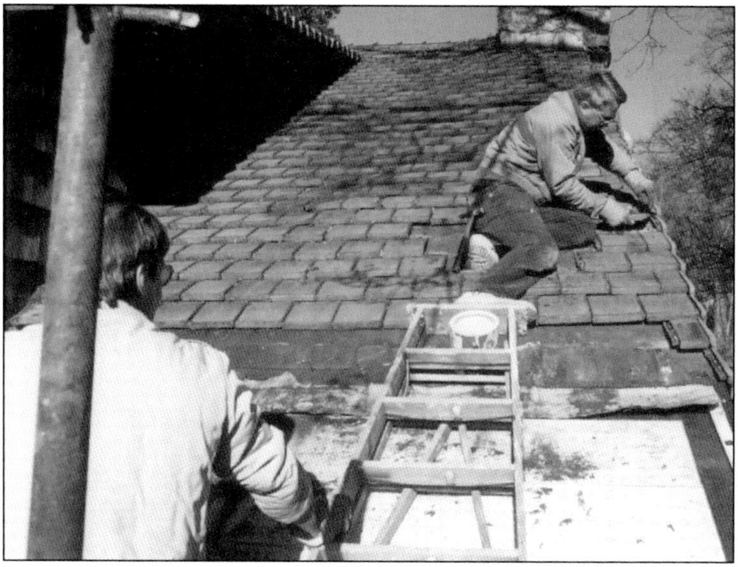

Workers reinstall the original green ceramic roof tiles on the northeast corner of the log house. The tiles had been carefully removed to allow for structural repairs below. Tiles that were broken or missing were replaced by identical new ones made by the original manufacturer, Ludowici.

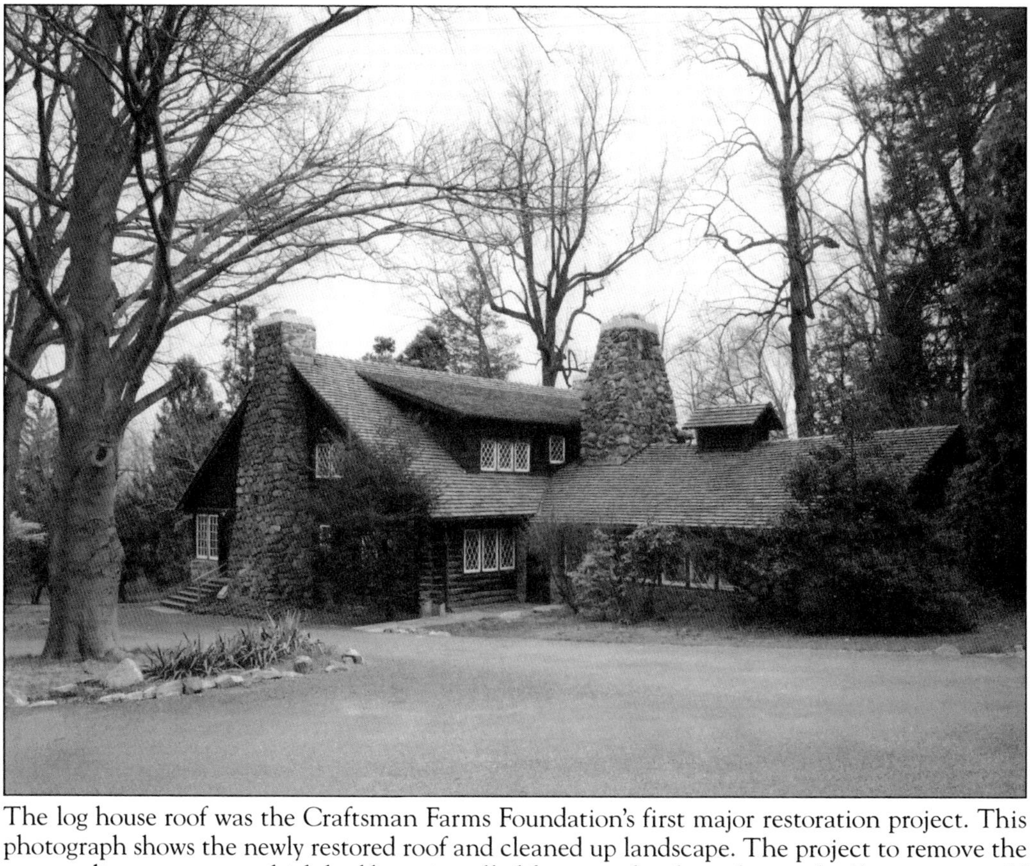

The log house roof was the Craftsman Farms Foundation's first major restoration project. This photograph shows the newly restored roof and cleaned up landscape. The project to remove the surrounding pavement, which had been installed for ease of parking during the Farny era, would wait for another day.

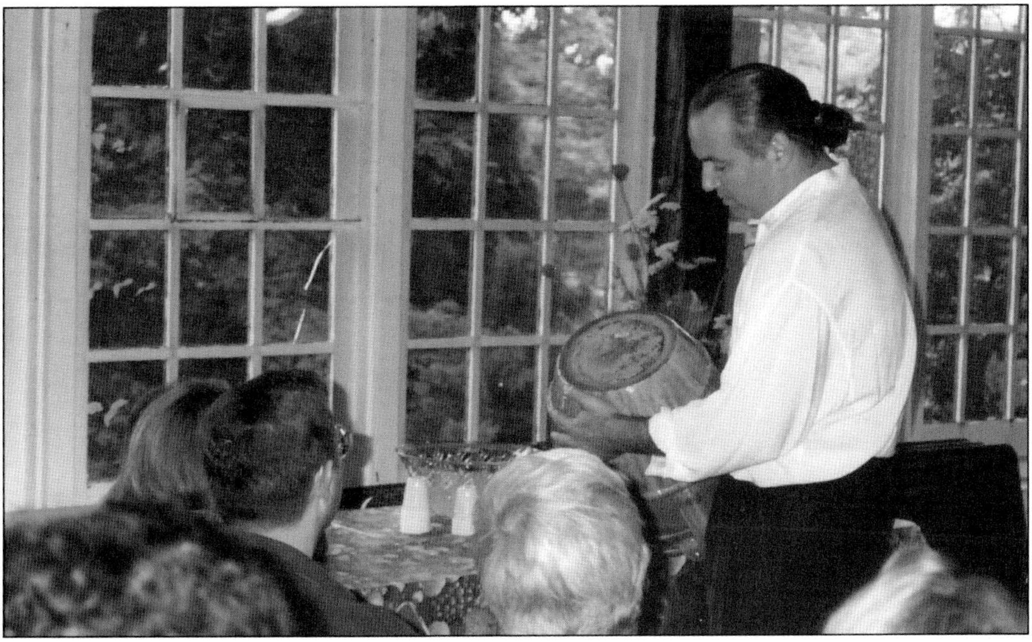

A year of great energy and excitement at Craftsman Farms was 1994. The Craftsman Farms Foundation was not content to simply work on necessary restoration projects. The board of trustees was anxious to serve the public through small exhibitions and educational programs even before the log house was restored. The top photograph shows David Rago, who is well known today as a pottery expert on *Antiques Roadshow*, lecturing on Fulper pottery at one of those 1994 programs. The bottom photograph shows another 1994 exhibition and program on the British Arts and Crafts giant Charles Robert Ashbee. It was curated by David Lowden (right). Ashbee's talented daughter, Felicity Ashbee (left), who was described in her obituary in the *Independent* as "the last close link with the inner circle of extraordinary creative talents fostered or inspired by William Morris," attended.

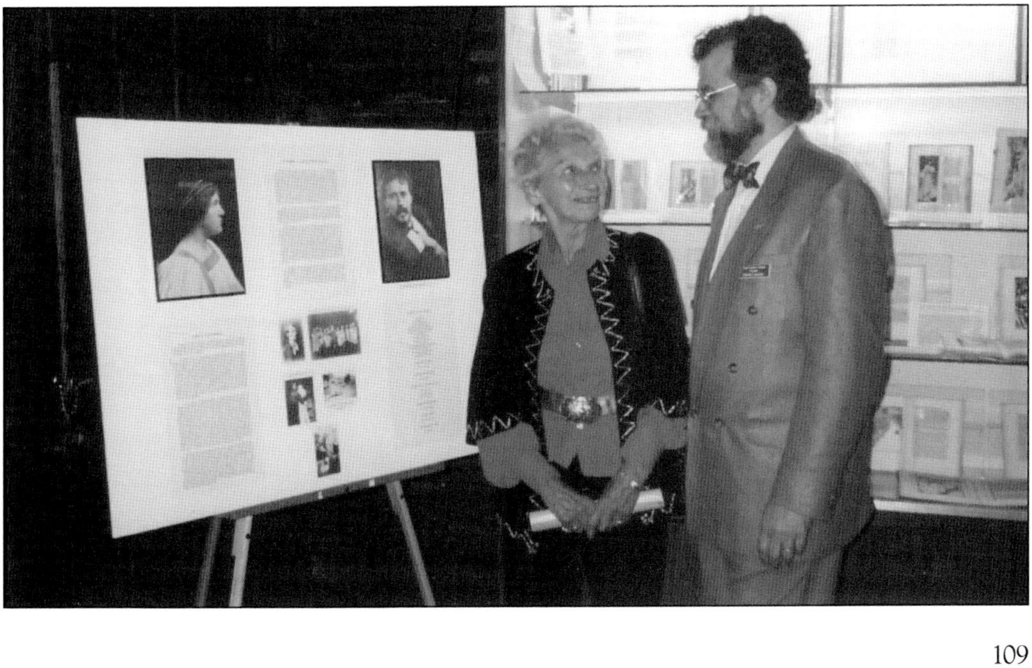

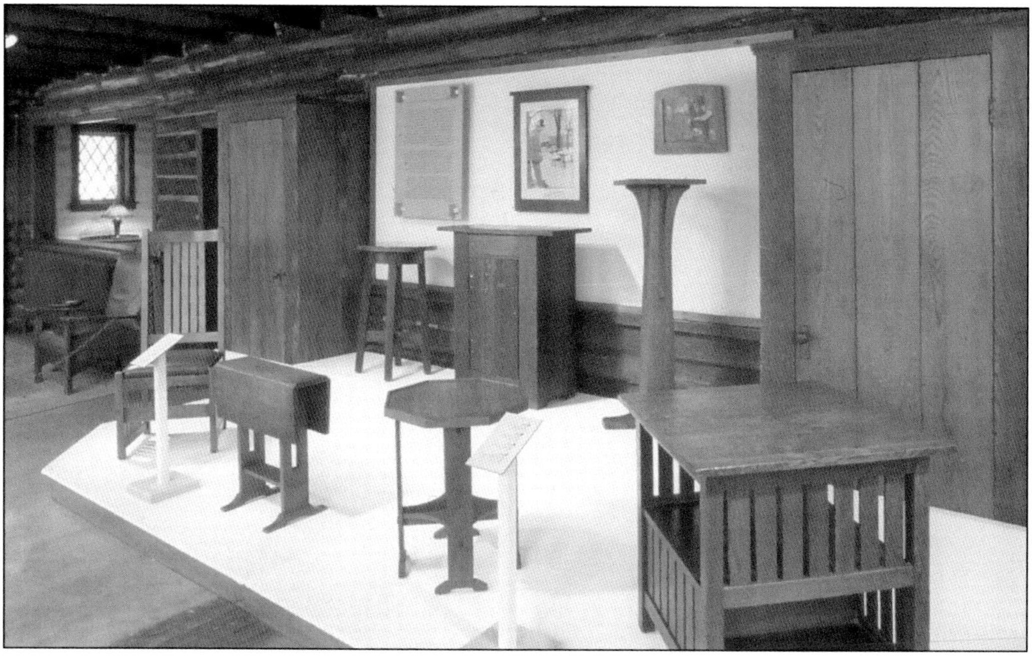

In 1996, the exhibition *Innovation and Derivation: The Contribution of L.& J.G. Stickley to the Arts and Crafts Movement* was presented in the log house. It was accompanied by a detailed catalogue written by exhibition curators Donald A. Davidoff and Stephen Gray. The photograph shows the log house porch with a portion of the exhibition.

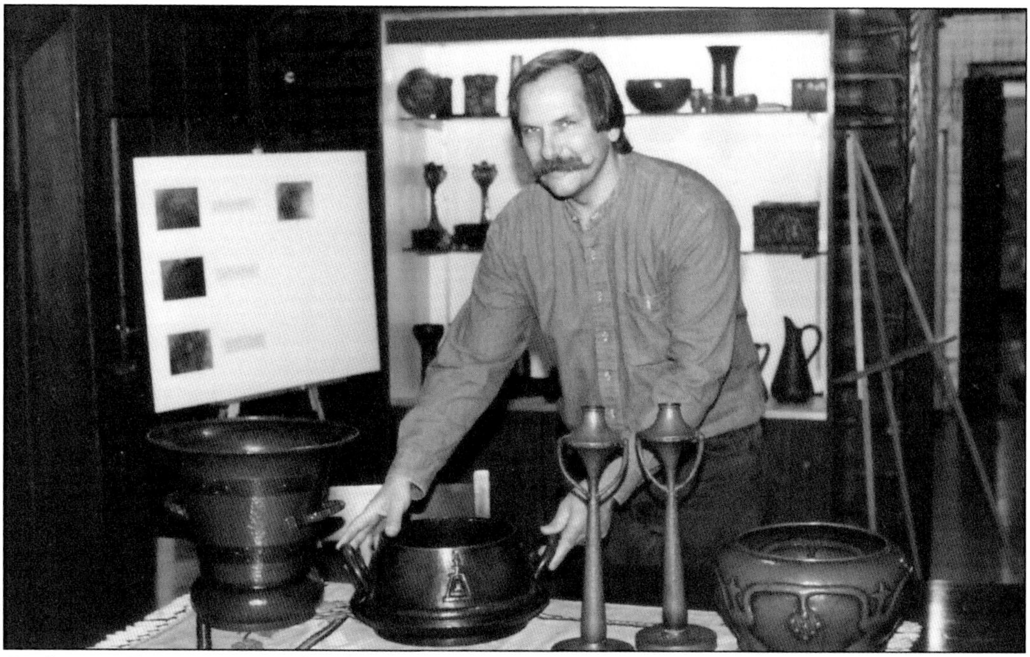

Craftsman Farms continued to host exhibitions and programs. In this photograph, Theodore Lytwyn is presenting a lecture in conjunction with the 1996 exhibition *By Hammer and Hand*. The metalwork exhibition was curated by Lytwyn and Raymond Groll and was a perfect opportunity for log house visitors to gain a better understanding of the house's five fireplace hoods.

The log house required major repairs and restoration throughout the entire structure. This 2004 photograph shows the dining room floor as it was being restored. Most of the white paint was removed from the walls, and the picture window was restored to its Stickley-era appearance.

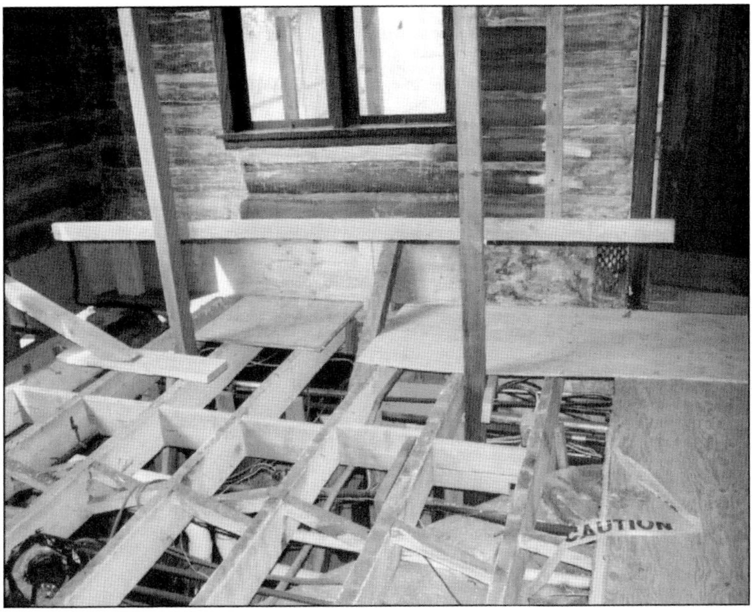


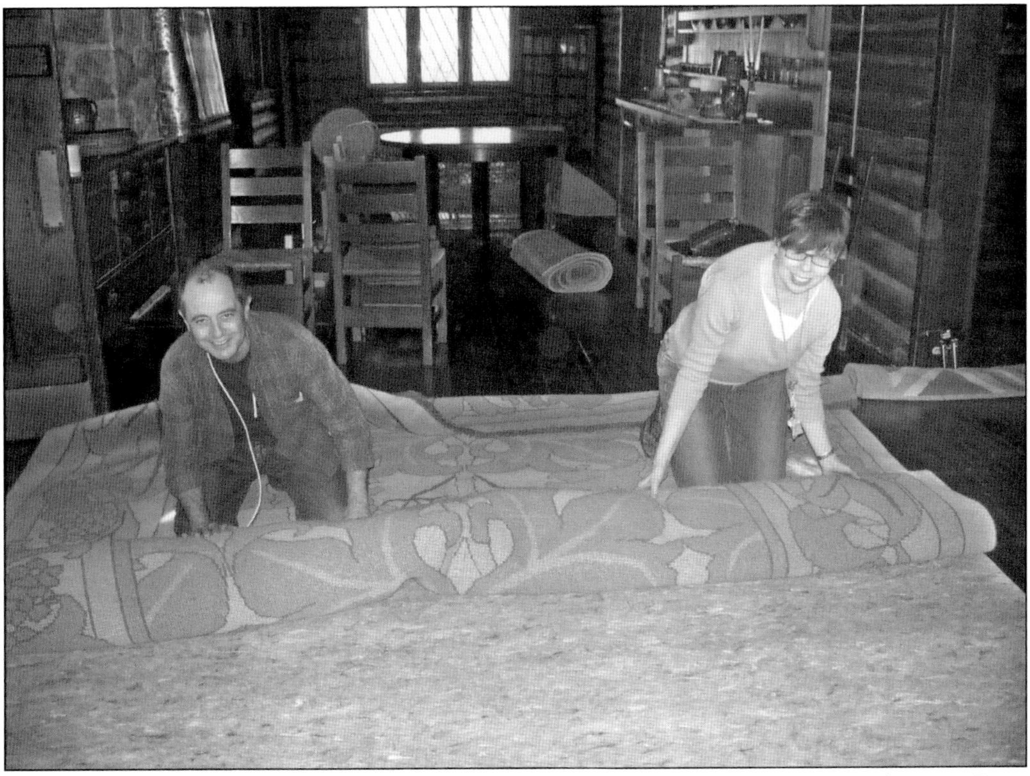

The dining room was brought even closer to its original appearance in 2006, when Jax Rugs provided hand-woven reproductions of the original Stickley carpets. Jax Rugs owner Del Martin (left) and Craftsman Farms Foundation executive director Arlette Klaric (right) are rolling out one of the three donated carpets in the dining room.

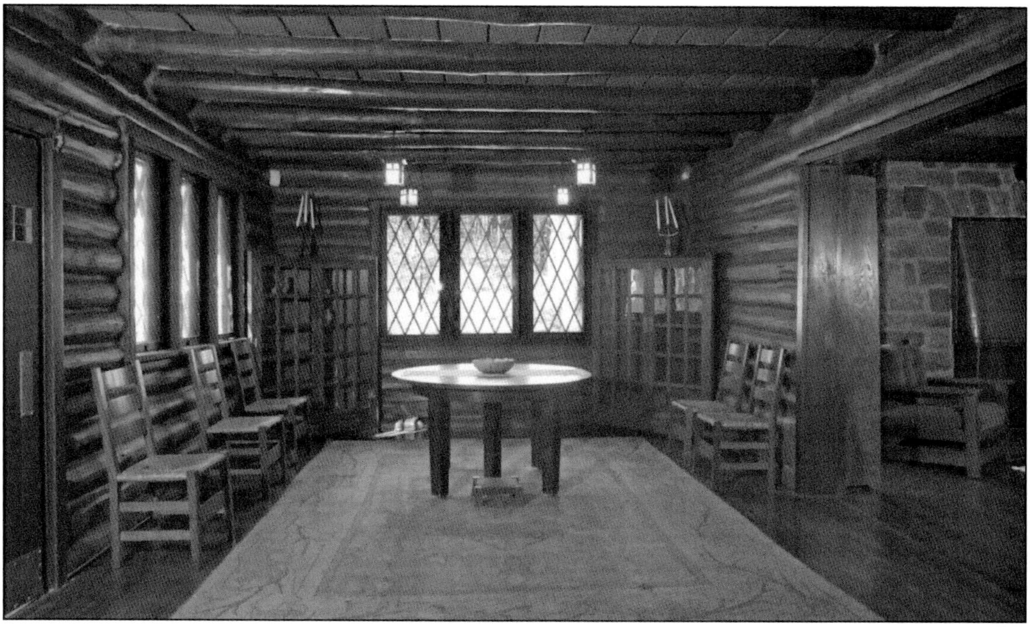

The corner cabinets, shown here in the north end of the dining room, were custom made just for this location by Gustav Stickley's craftsmen. The Craftsman Farms Foundation ran a fundraising campaign called "I'm in Craftsman Farms' Corner" to successfully purchase them for $125,000 at a Christie's auction. These authentic furnishings combine to return the room to its 1911 appearance.

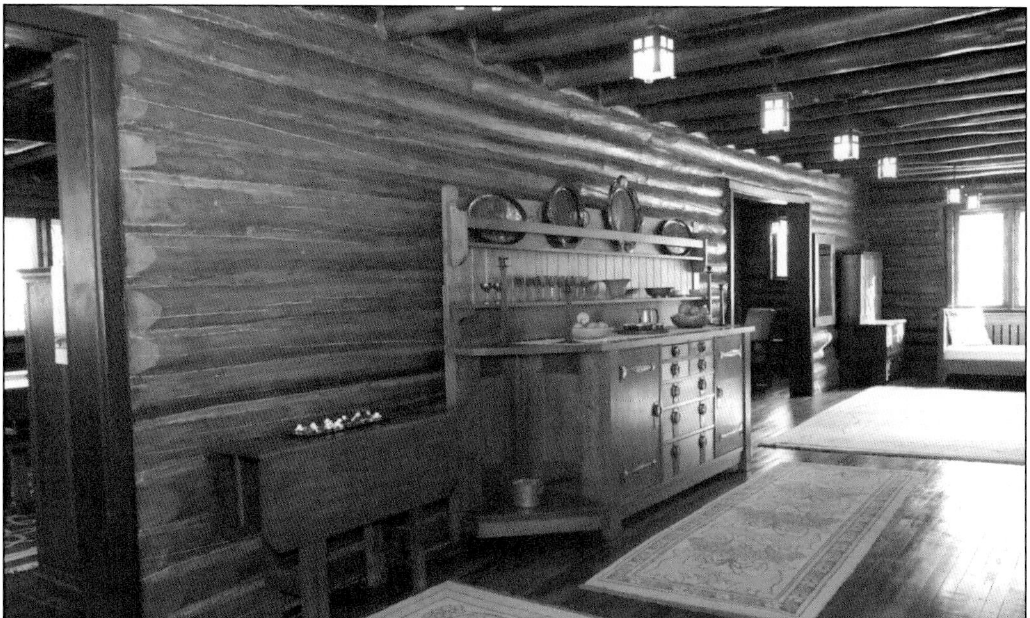

This very rare Stickley sideboard was restored and donated by an anonymous donor in 1994. This sideboard, the three reproduction carpets, the installation of the lighting, the restoration of the logs, and the addition of smaller objects all help make the dining room look very much like it did when Gustav Stickley lived here with his family.

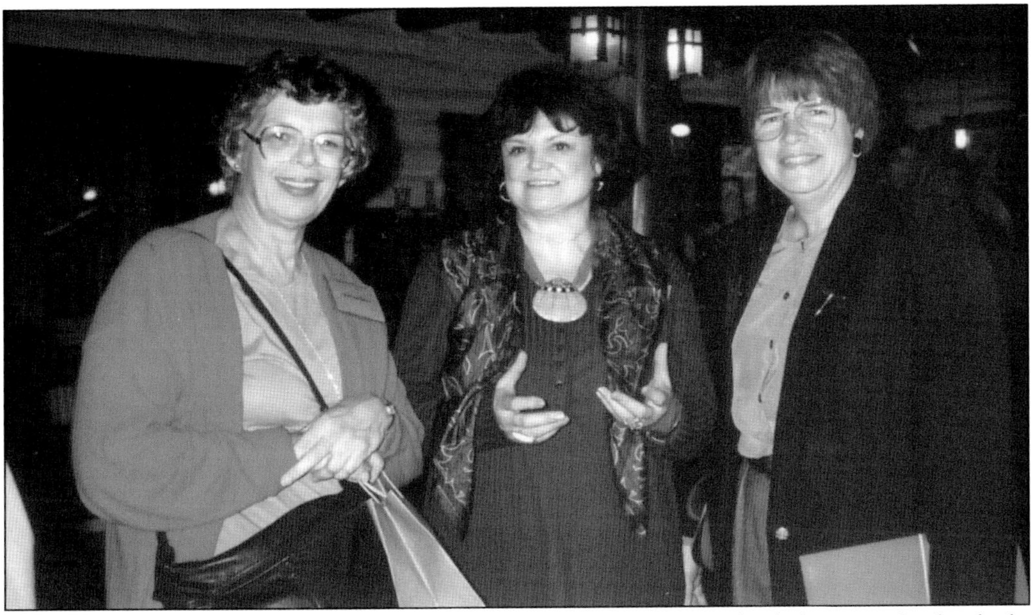

The successful rescue and rebirth of Craftsman Farm is largely due to the dedication of individuals. Three individuals who have made a difference are shown in the log house in this 1994 photograph. From left to right are the author of *Gustav Stickley, the Craftsman*, MaryAnn Smith; Roycroft champion and campus historian Edythe "Kitty" Turgeon; and Gustav Stickley's great-granddaughter, Barbara Fuldner.

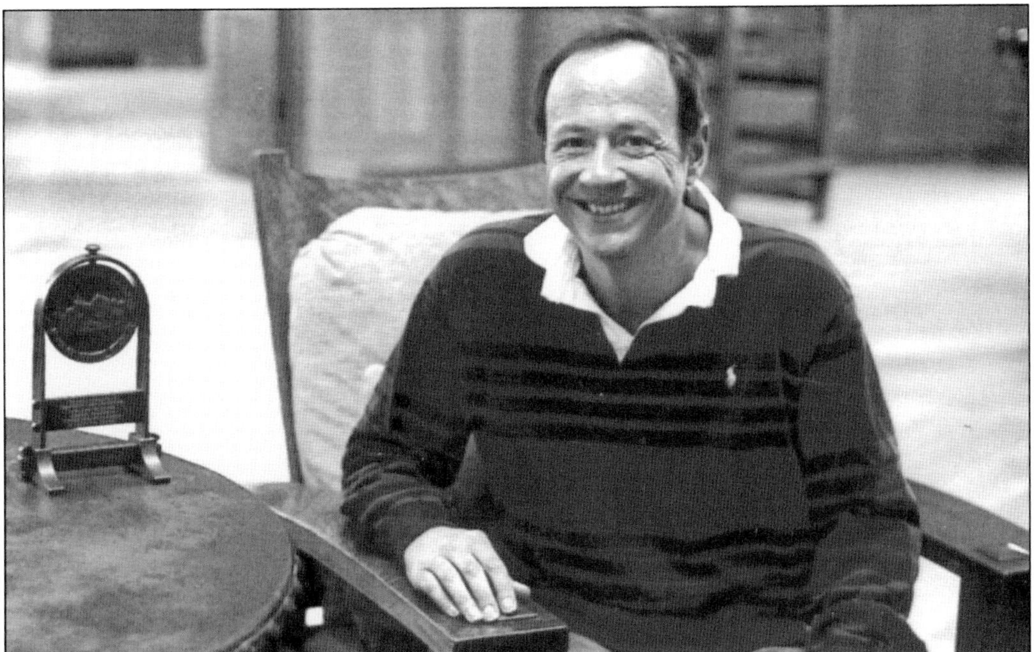

In 1993, twenty pieces of Stickley's furniture were returned to Craftsman Farms through generous efforts of the noted Arts and Crafts collector Paul Fiore and an anonymous benefactor. This 1995 photograph shows Fiore when he received the Craftsman Farms Foundation's first Als Ik Kan Award.

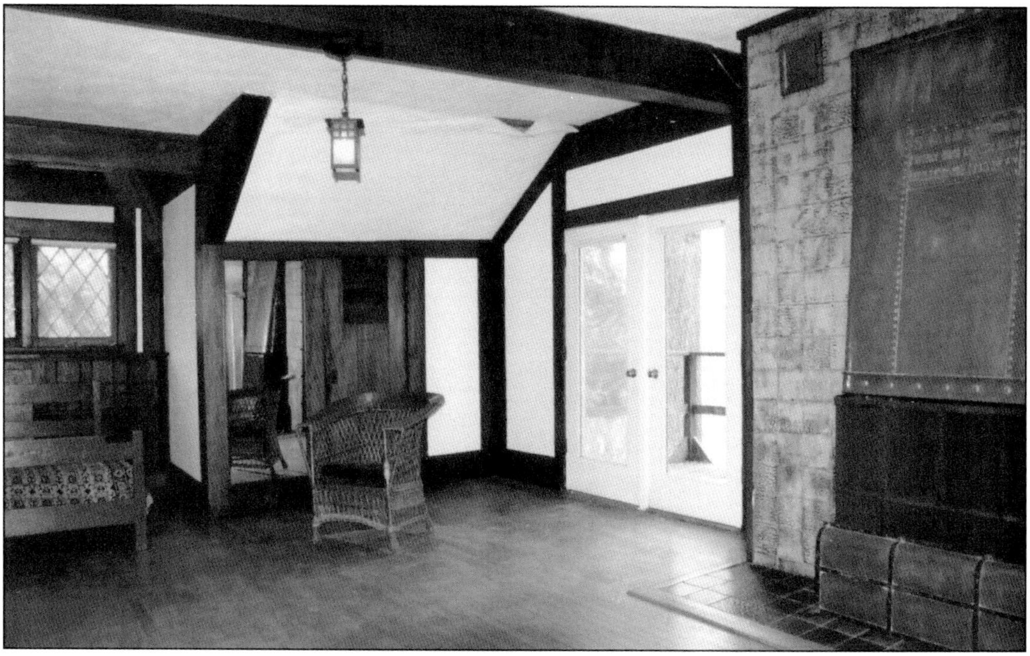

This 1993 photograph shows the girls' bedroom interpreted for the public with some of the original furnishings that had been donated by a benefactor. On the right are the glass doors to the balcony added by the Farnys. These were later removed and the windows restored.

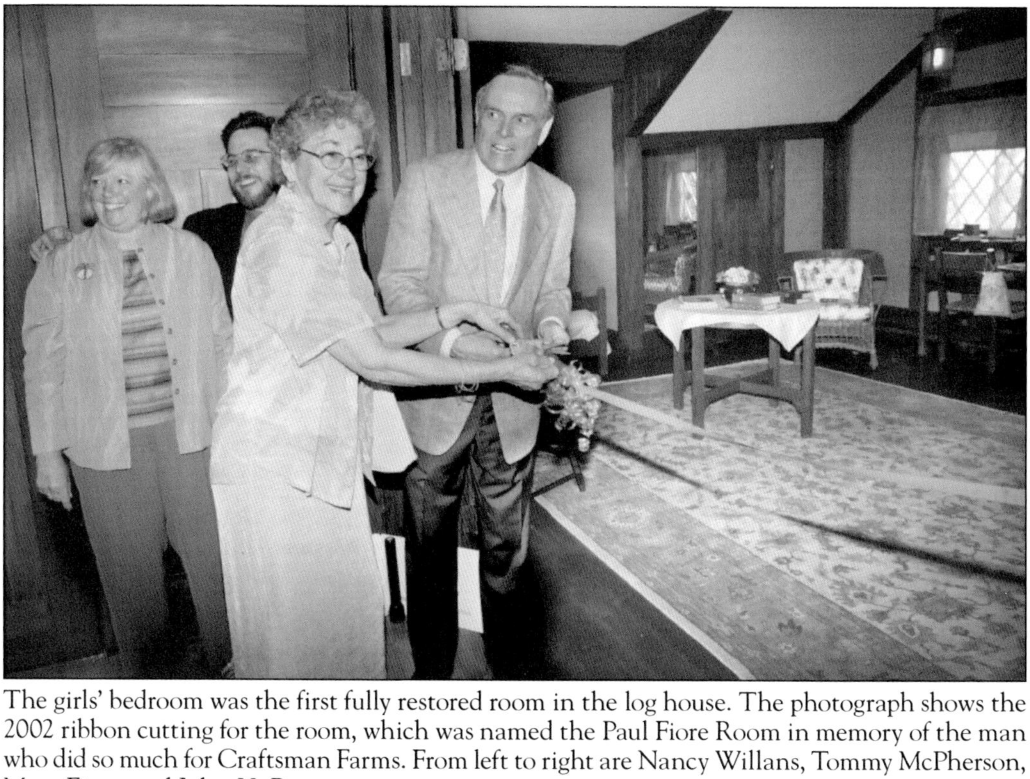

The girls' bedroom was the first fully restored room in the log house. The photograph shows the 2002 ribbon cutting for the room, which was named the Paul Fiore Room in memory of the man who did so much for Craftsman Farms. From left to right are Nancy Willans, Tommy McPherson, Mary Fiore, and John H. Bryan.

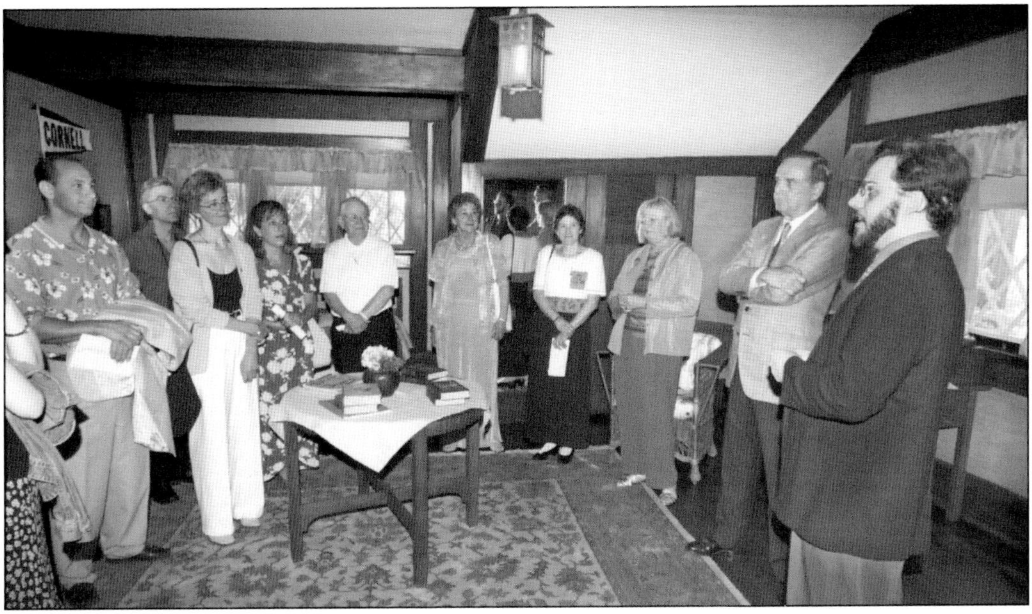


Executive director Tommy McPherson (far right) makes remarks at the celebration of the completion of the girls' bedroom. The fully restored, diamond-paned casement windows are in place where the Farny family had installed a balcony with large glass doors. The windows can be seen behind McPherson.

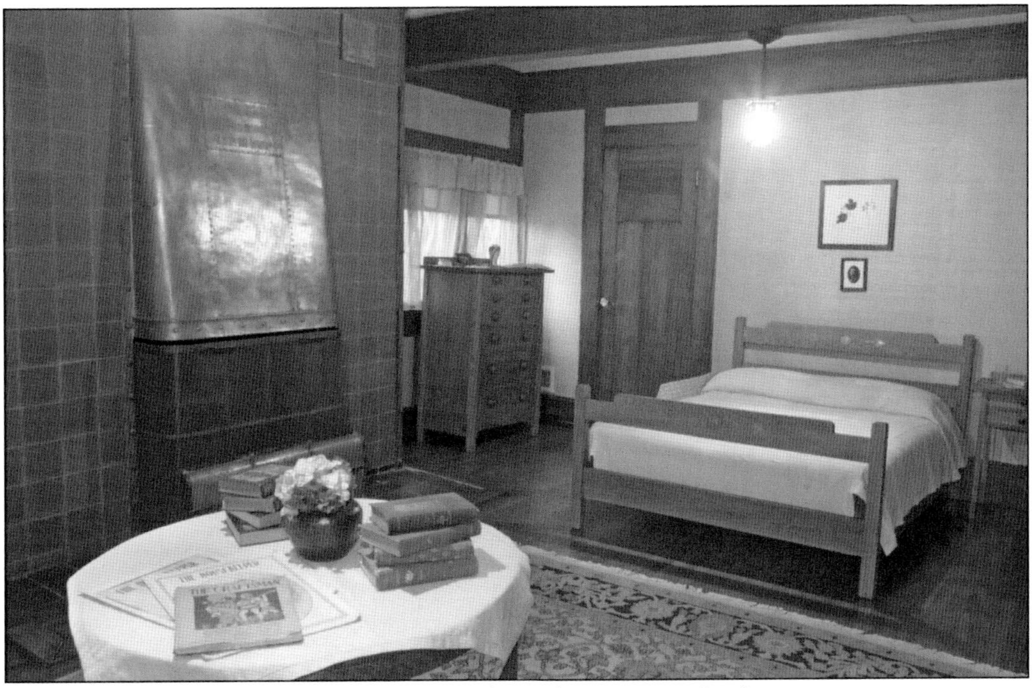

This contemporary view of the girls' bedroom shows the great care that has gone into its restoration. The silver-grey grass-cloth walls and low-luster wood trim look just as they did 100 years ago. The hammered brass fireplace hood and the Grueby tiles that surround the fireplace are highlights of the room.

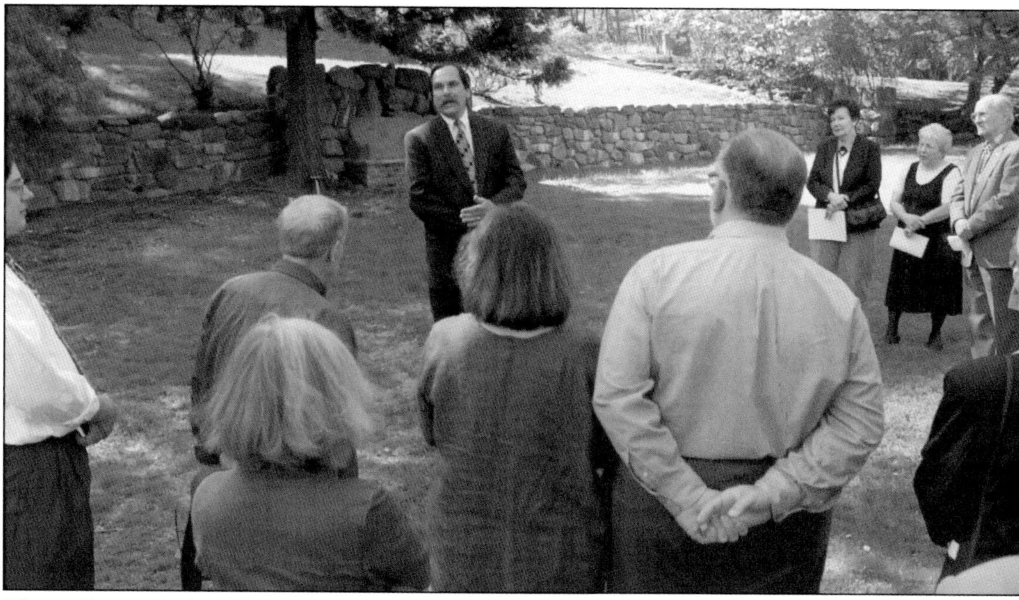

The wall surrounding the sunken area known as the rose garden had deteriorated over time. Using photographs as a reference, the wall was carefully rebuilt by skilled stone workers. In this 2000 photograph, board president Theodore Lytwyn (center) addresses the crowd at the restored wall and dedicates the bench in memory of former executive director Nancy Strathearn.

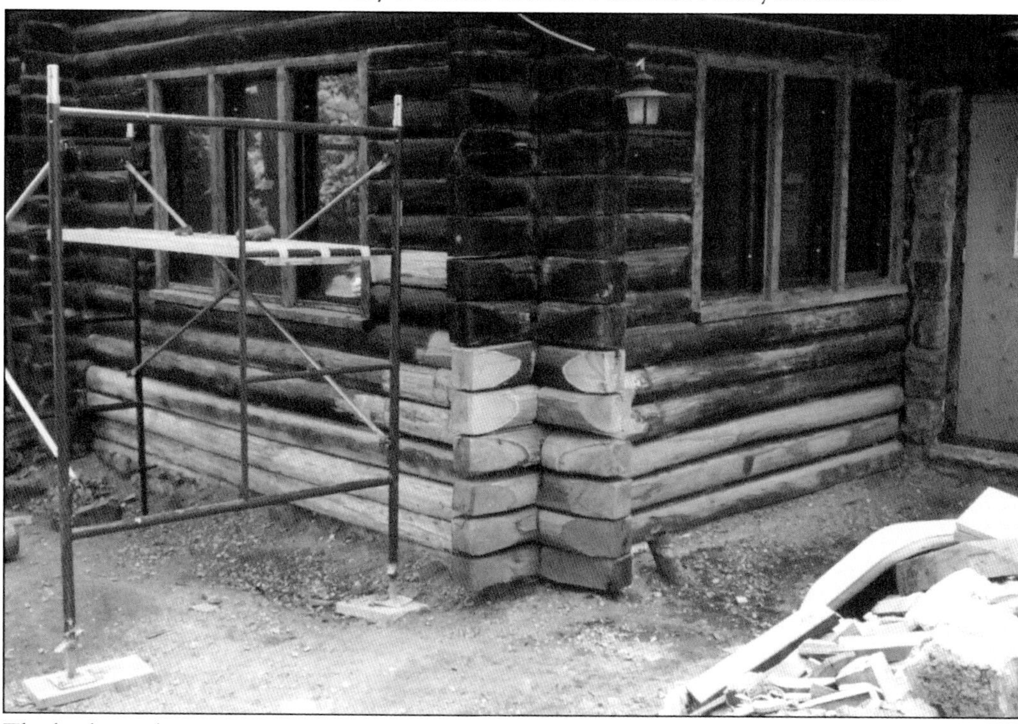

The log house has required major structural repairs since the Craftsman Farms Foundation began its work. Steel support structures have been used to hold the weight of the roof, support the staircase, and even inside door frames. All have been completely hidden from view. In this 2004 photograph, some rotted logs in the corner of the dining room are being replaced.

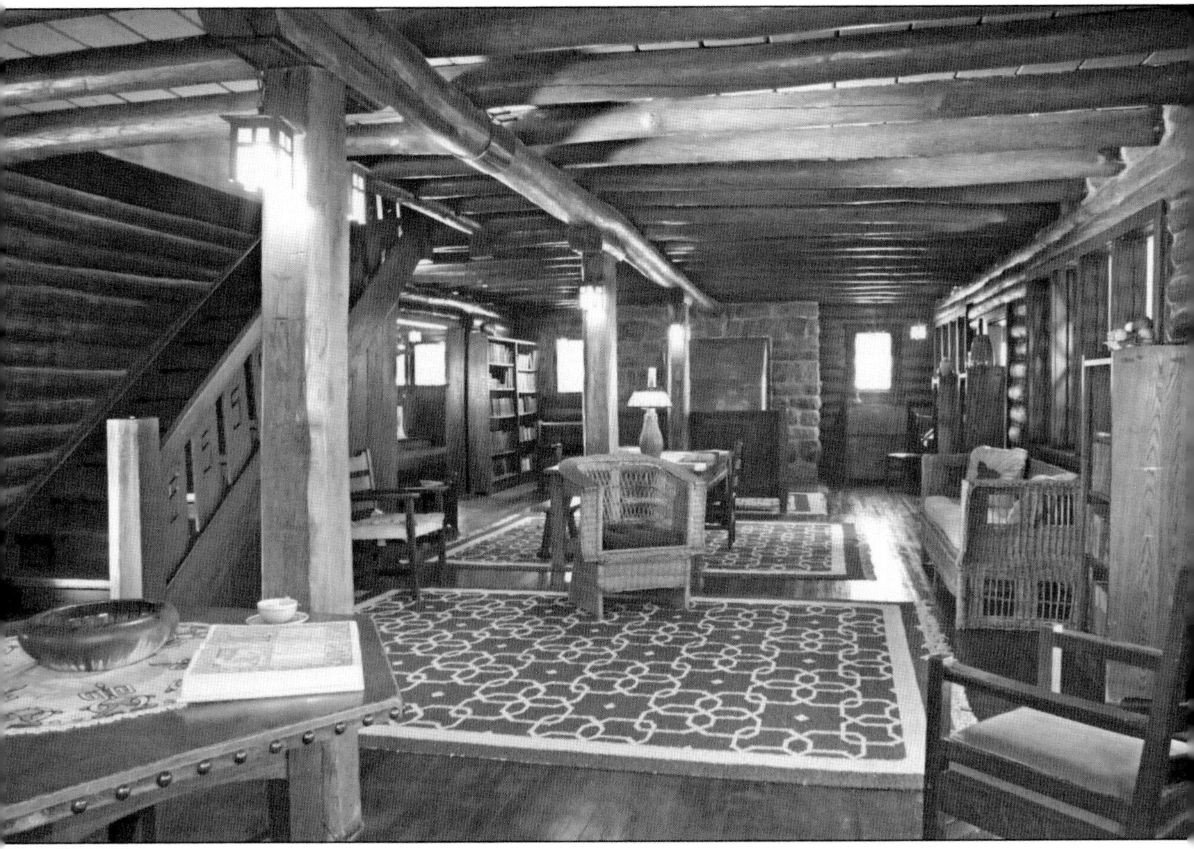

This contemporary photograph shows the fully restored living room looking to the north. The fieldstones that surround the fireplace at the back of the photograph are no longer white, the lanterns have been installed, and the room offers the tranquil welcome of earlier days. Over time, many of the original furnishings were returned to the log house, either by loan, gift, or purchase. Additional pieces of similar form to the original were also added to the rooms. Stickley's period photographs were used for reference, with a goal of interpreting the log house for visitors in the most authentic way possible. By 1997, the Craftsman Farms Foundation had formally adopted the name of the Stickley Museum at Craftsman Farms and began using it wherever possible. With three rooms largely restored, the log house began operating as a historic house museum and was open to visitors from April though November.

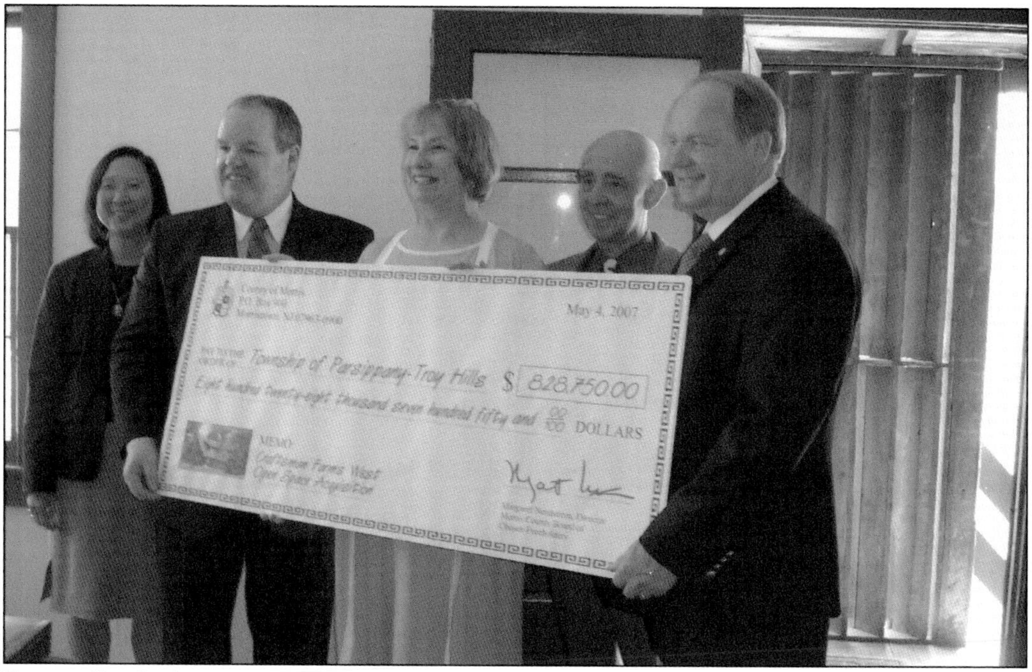


Eventually, the last remaining Craftsman Farms acreage came up for sale. This four-acre parcel of land contained the remaining Stickley farm buildings. Though cash-strapped, the Craftsman Farms Foundation purchased the property in 1999 to prevent its further development. But the high costs of monthly mortgage payments forced the self-funded foundation to reduce its programs and historic preservation. Thankfully, in the spring of 2007, the Township of Parsippany–Troy Hills obtained Open Space funds to purchase the property. The foundation now holds a 25-year lease on all 30 acres of the remaining property. Presenting the check from left to right are township administrator Jasmine Lim, Mayor Michael Luther, freeholder Margaret Nordstrom, freeholder Jack Schrier, and freeholder Douglas Cabana. At the horse barn during the same event are, from left to right, freeholder Gene F. Feyl, Councilman Michael dePierro, freeholder Jack Schrier, executive director Heather Stivison, freeholder Douglas Cabana, freeholder Margaret Nordstrom, Mayor Michael Luther, board treasurer Robert Burchell, trustee emerita Nancy Willans, board president Davey Willans, board secretary Martha Weinstein, trustee Karen Slifkin, and board vice president Laura Reilly.

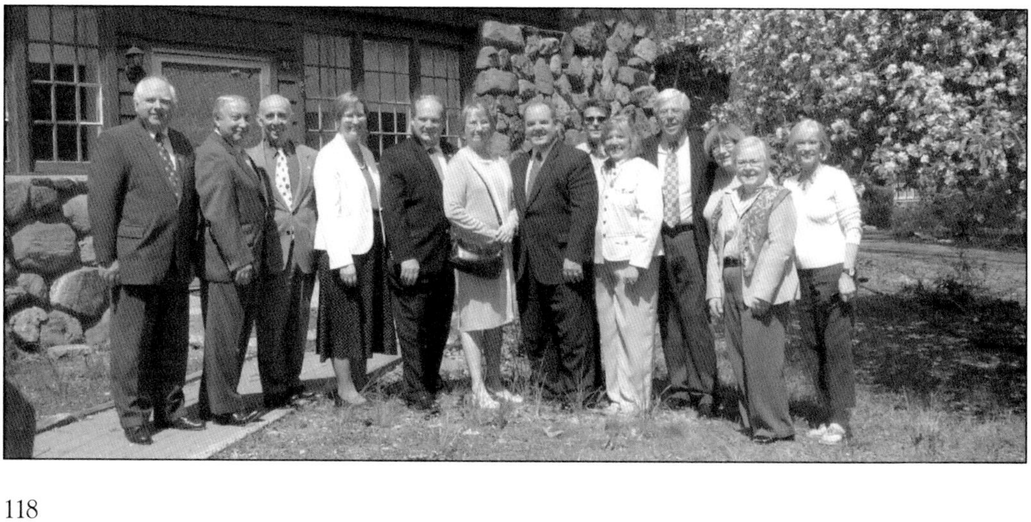

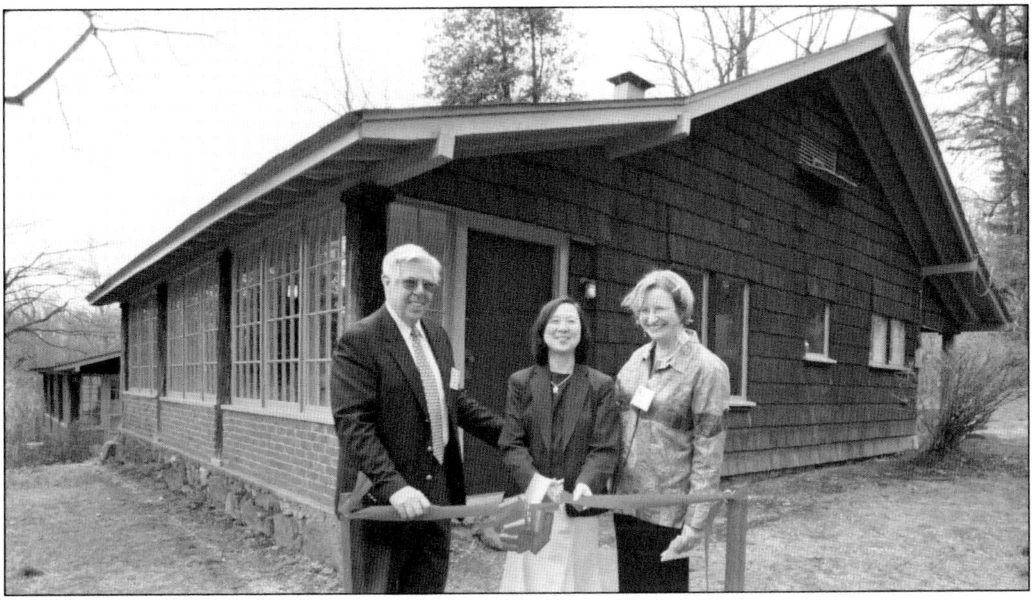

Freed up from the burden of mortgage payments, the Craftsman Farms Foundation resumed its core mission of restoration and programming. The museum commissioned a full historic site master plan and began to undertake many new projects, going well beyond the log house. The Farny-era annex was rehabilitated and converted into an education building. The twin cottages had their roofs fully restored. The north cottage was rehabilitated inside. At the ribbon cutting for the newly opened north cottage are, from left to right, board president Davey Willans; Jasmine Lim, administrator for the Township of Parsippany–Troy Hills; and executive director Heather Stivison.

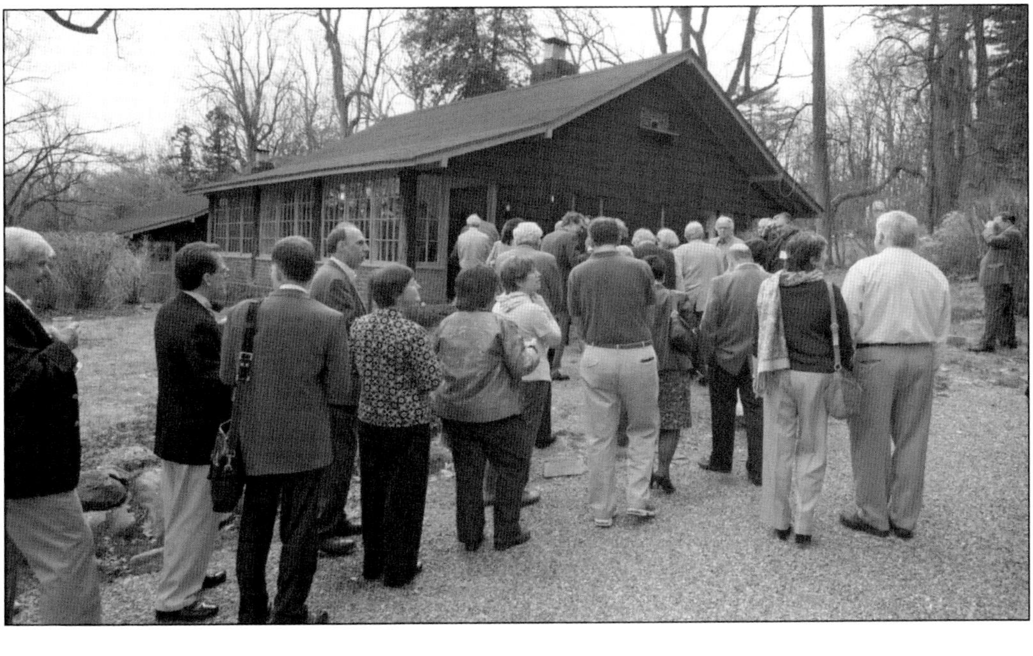

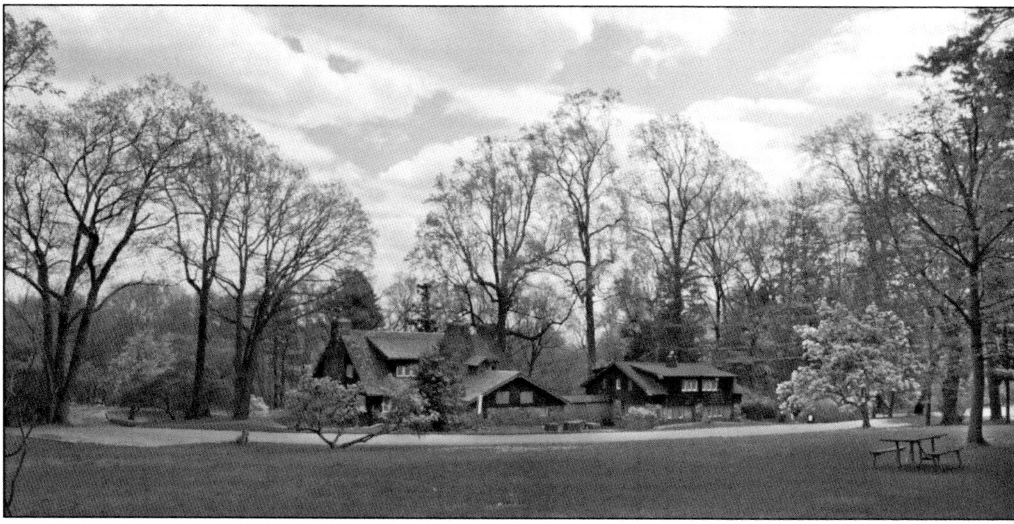

This 2010 photograph is strikingly reminiscent of the 1920 panorama shown on pages 30–31. The swimming pool and blacktop parking have been removed, the grounds have been landscaped, and much of the central property has been returned to its original appearance.

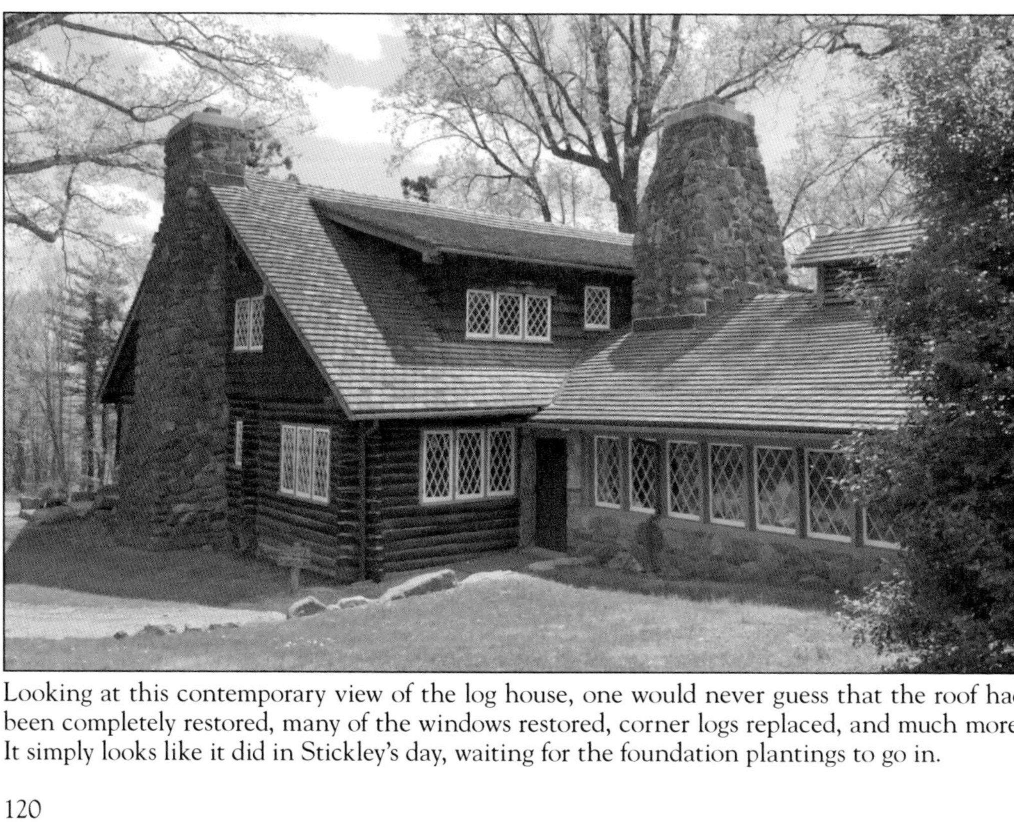

Looking at this contemporary view of the log house, one would never guess that the roof had been completely restored, many of the windows restored, corner logs replaced, and much more. It simply looks like it did in Stickley's day, waiting for the foundation plantings to go in.

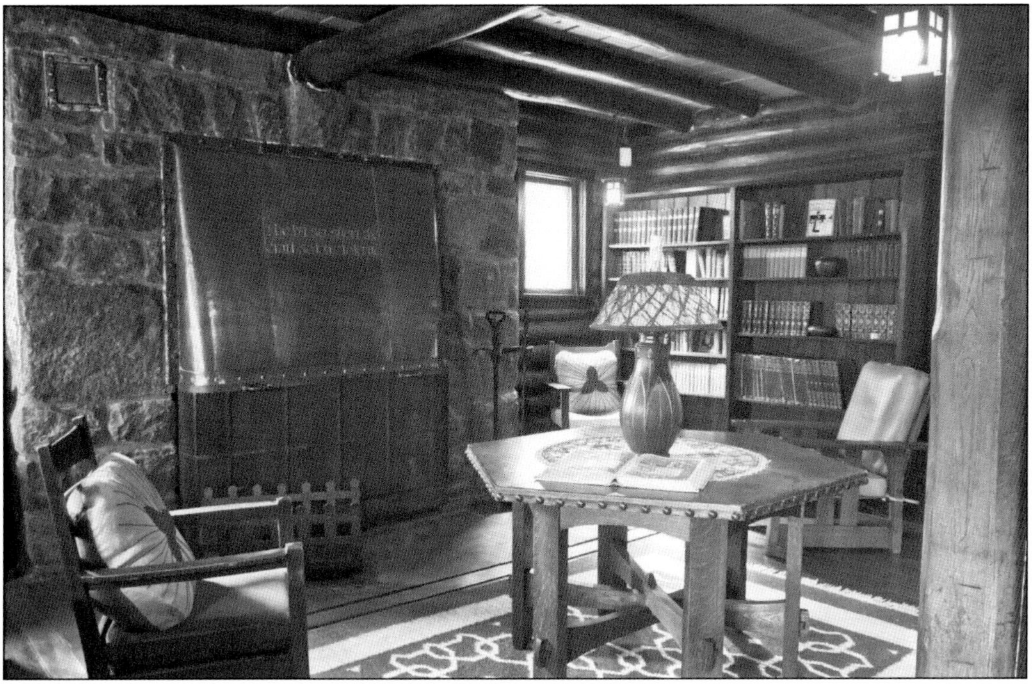

The cozy reading nook at the south end of the living room today looks virtually identical to the way it appeared in 1911. The bookcase is the actual chestnut bookcase originally made for the log house. The Grueby lamp base on the hexagonal table is also original.

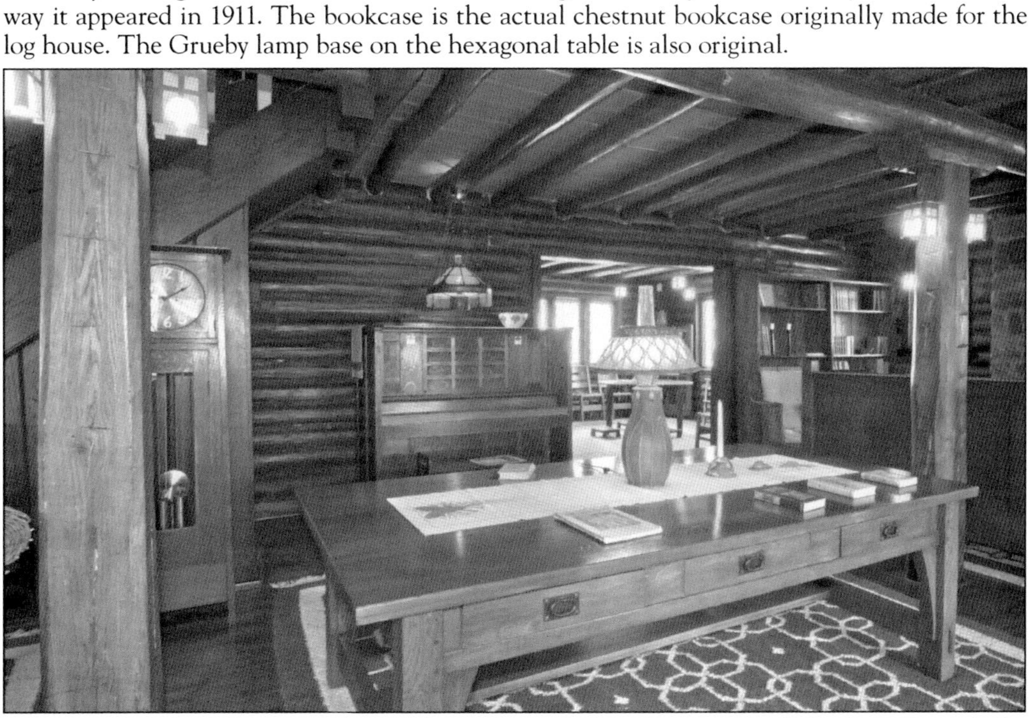

This view is of the fully restored central portion of the living room. The light glows through the mica shade of the piano lamp positioned over the upright piano. The elm library table in front has been restored to its green stain. The tall case clock visible on the left was on loan at the time this photograph was taken.

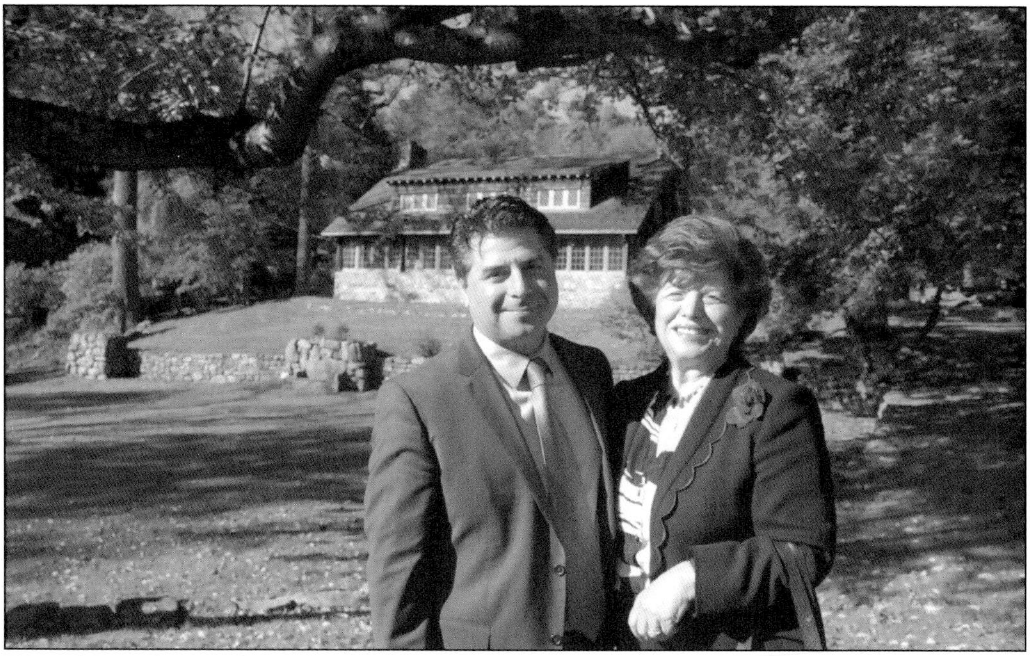

Edward J. Audi, president of L. & J.G. Stickley, Inc., poses with Aminy Audi, chair and chief executive officer, in front of the log house during the 100th anniversary weekend. The Audis have been dedicated supporters of Craftsman Farms, and Aminy Audi was an honorary chair of the anniversary celebration.

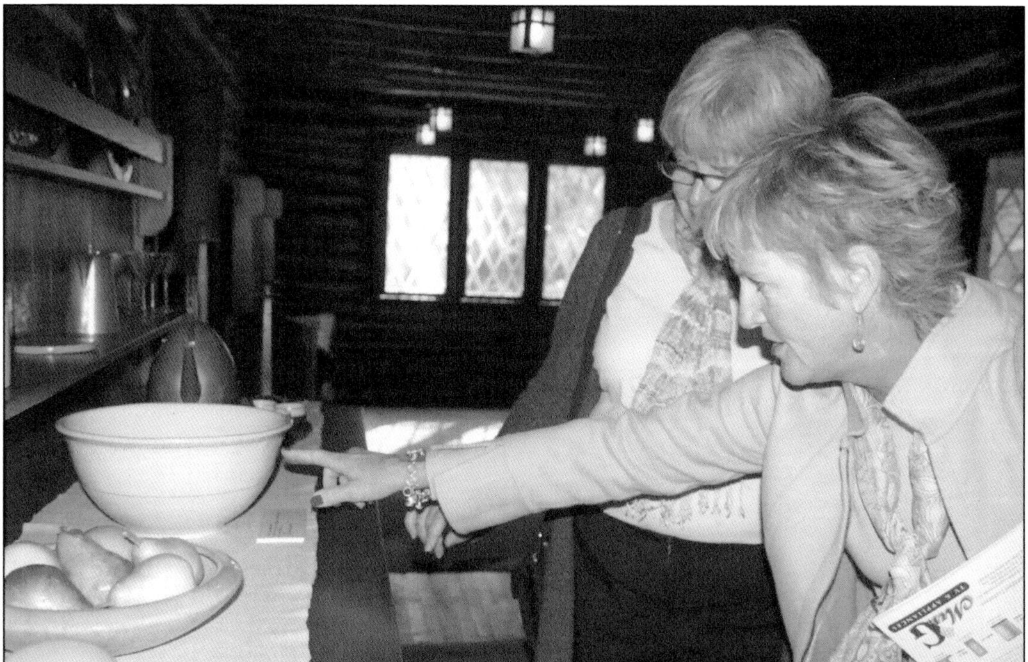

Stickley descendants Barbara Fuldner (back) and Cynthia McGinn (front) examine a bowl that once belonged to their great-grandfather Gustav Stickley. Barbara is the granddaughter of Barbara Stickley Wiles. Cynthia is the granddaughter of Mildred Stickley Cruess.

In 2013, the Stickley Museum at Craftsman Farms expanded its regular hours, which had been seasonal. It is now open to the public five days a week all year round. Daily attendance continues to grow. In this 2011 photograph, a group arrives for a docented house tour.

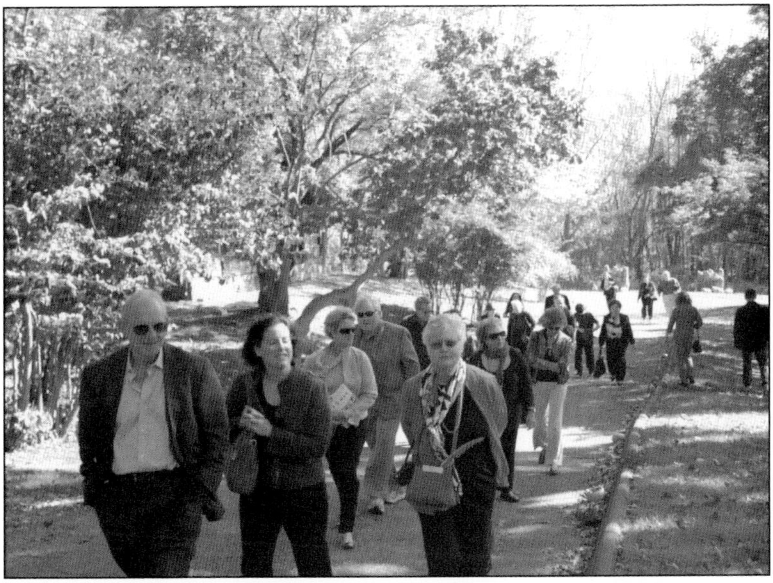

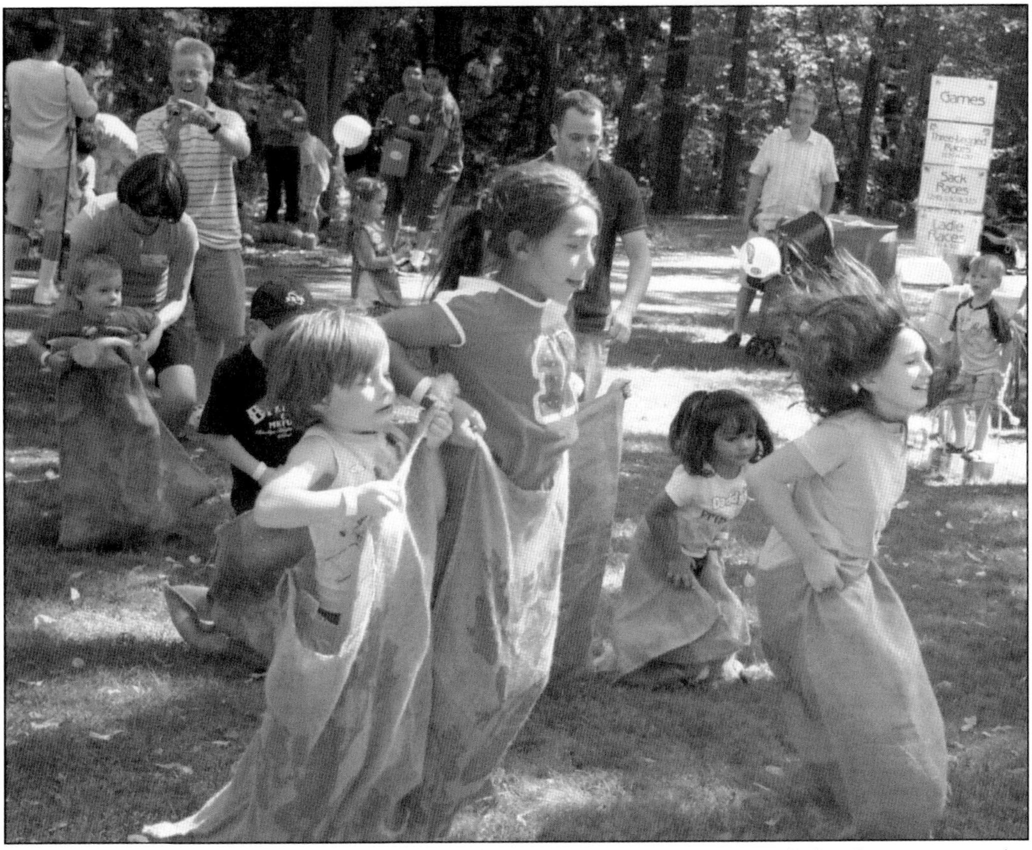


Beginning with a fall family day in 2007, the museum's mission expanded to better serve the community and to reach all ages. Hundreds of children participate in programs each year that encourage an understanding of history through experiences. Here, children explore the old-fashioned fun of a sack race during a family day on the grounds.

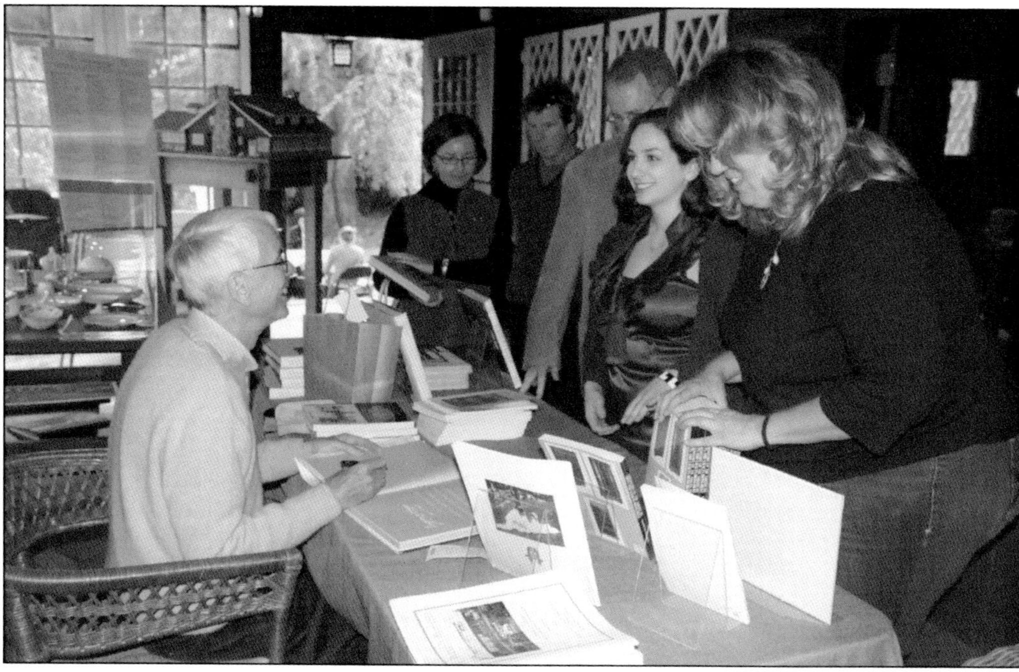

The museum offers an array of adult programming, including lectures by scholars, workshops led by artisans, and special programs on a wide variety of topics. Here, Stickley scholar David Cathers is autographing books on the log house porch. Three unidentified visitors wait in line while Kristen McCauley (looking down at Cathers) and Betty Wyka (right) get their books signed.

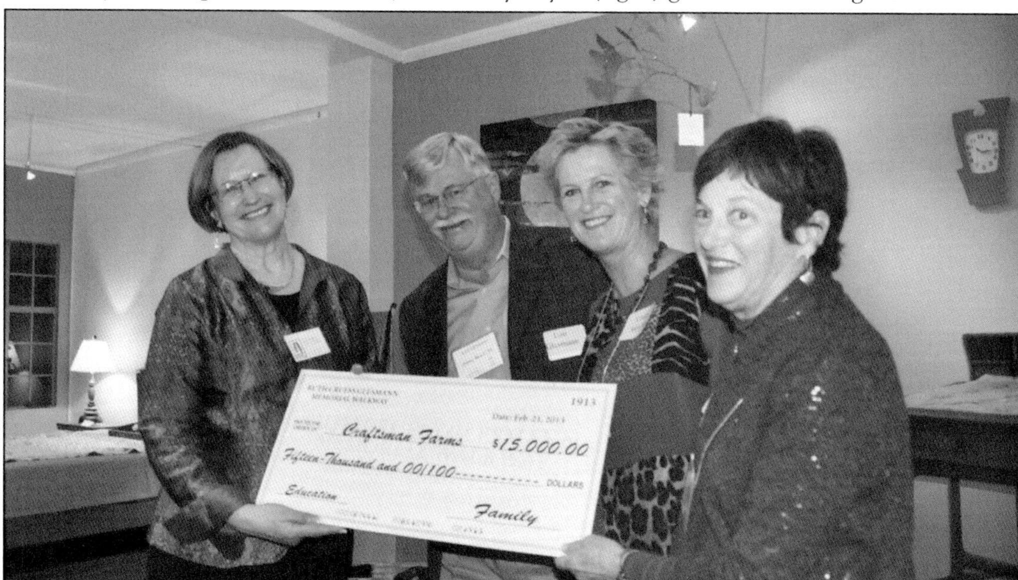


Three Stickley descendants made this gift in memory of their mother, Ruth Cruess Glesmann, who passed away in 2012. She was Gustav Stickley's granddaughter and lived at Craftsman Farms as a child, so it is especially appropriate that the walkway to the education building has been named in her honor. From left to right are executive director Heather Stivison, Louis Glesmann, Cynthia McGinn, and Nancy Calderwood.

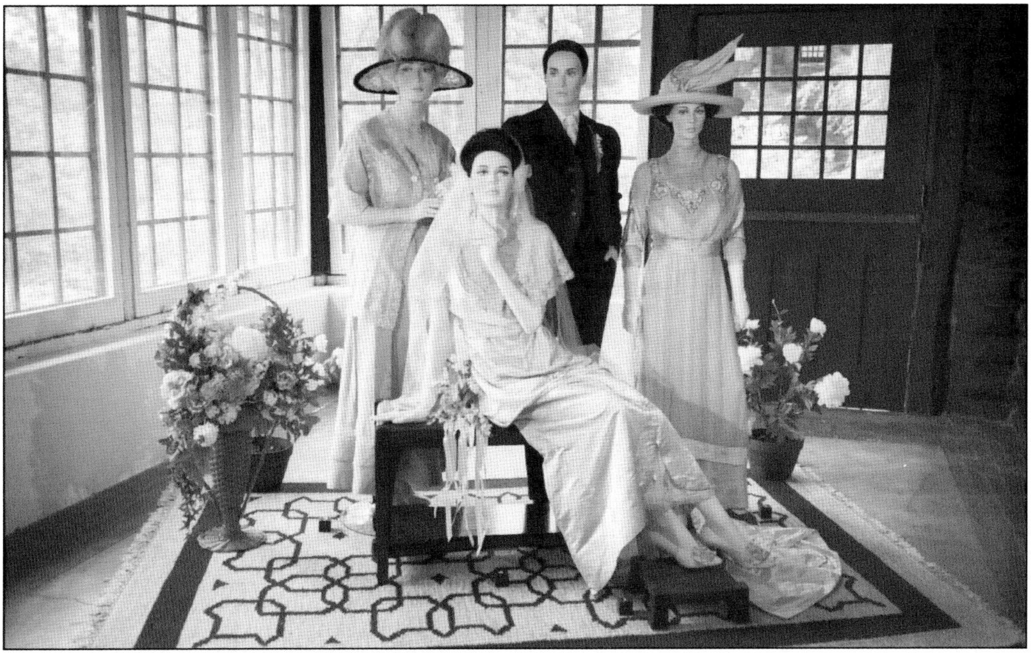

The Stickley Museum partnered with Syracuse University on the exhibition *Styling an American Family: The 1910s at Gustav Stickley's Craftsman Farms*, which ran from September 8, 2012, to January 7, 2013. Curator Jeffrey Mayer selected the garments from the university's Sue Ann Genet Costume Collection. The exhibition featured 34 mannequins in vignettes throughout the house. This wedding vignette was on the porch. (Courtesy Stephen Sartori, the Stickley Museum at Craftsman Farms.)

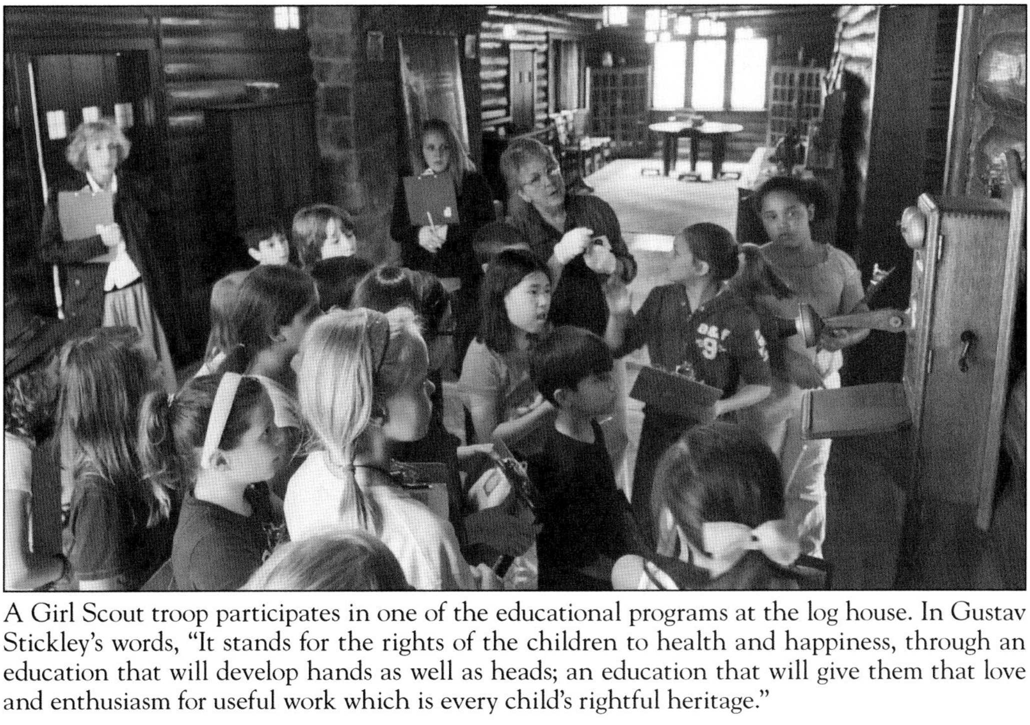

A Girl Scout troop participates in one of the educational programs at the log house. In Gustav Stickley's words, "It stands for the rights of the children to health and happiness, through an education that will develop hands as well as heads; an education that will give them that love and enthusiasm for useful work which is every child's rightful heritage."

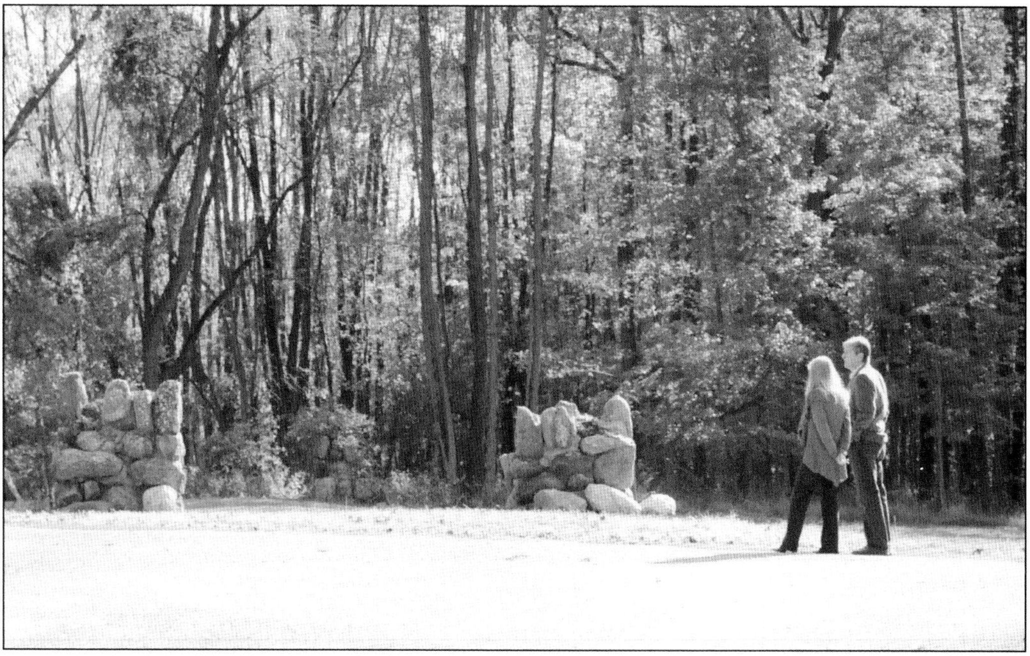

Bruce Johnson, director of the National Arts & Crafts Conference and Antiques Show at the Grove Park Inn since 1988, has been a longtime friend of Craftsman Farms. The museum has also been the recipient of several of his research grants. In this 2011 photograph, he and his wife, Leigh Ann, gaze wistfully at the stone piers, which have deteriorated significantly over time.

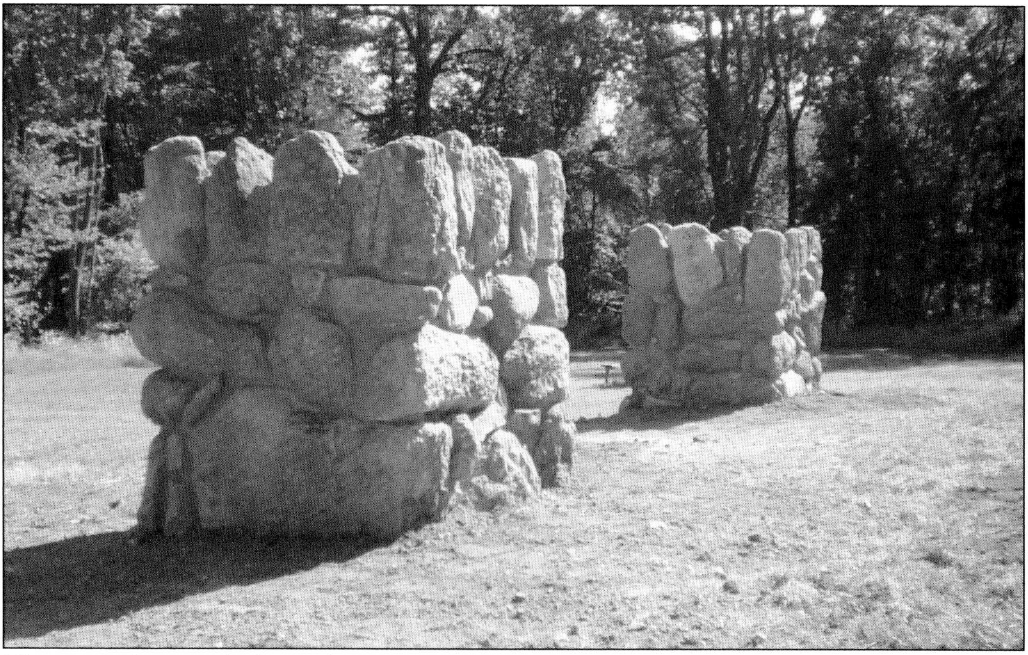

This 2013 photograph shows the same piers as the image above. A successful "Save the Stone Piers" campaign was announced in January 2013. By the fall of 2013, the project was underway. It will be filled with plantings to resemble the original photograph shown on page 28, and the restoration of Craftsman Farms will continue in the years to come.

# Bibliography


Bartinique, Patricia A., ed. *Gustav Stickley: His Craft*. Parsippany, NJ: The Craftsman Farms Foundation, Inc., 1992.

Cathers, David. *Gustav Stickley*. New York: Phaidon Press, 2003.

Cathers, David, ed. *Gustav Stickley's Craftsman Farms—A Pictorial History*. Rev. ed. Parsippany, NJ: The Craftsman Farms Foundation, Inc., 2009.

Cathers, David, and Alexander Vertikoff. *Stickley Style—Arts and Crafts Homes in the Craftsman Tradition*. New York: Simon and Schuster, 1999.

*The Craftsman* (1901–1916), http://digicoll.library.wisc.edu/cgi-bin/DLDecArts/DLDecArts-idx?type=browse&scope=DLDECARTS.HOMEDESIGN.

Hewitt, Mark Alan. *Gustave Stickley's Craftsman Farms: The Quest for an Arts and Crafts Utopia*. Syracuse: Syracuse University Press, 2001.

Smith, Mary Ann. *Gustav Stickley, the Craftsman*. Syracuse: Syracuse University Press, 1983. Reprint, New York: Dover Publications, 1992.

Tucker, Kevin W., et al. *Gustav Stickley and the American Arts & Crafts Movement*. Dallas: Dallas Museum of Art; New Haven: Yale University Press, 2010.